'BRIGHTWORK'

Traditional Paintwork on Leeds & Liverpool Ca

CW01667185

Mike Clarke & Sam Yates

Milepost Research 2009

© Copyright in this book is held by the authors. The terms of the Copyright, Designs & Patents Act, 1988, and the Duration of Copyright & Rights in Performance Regulations, 1995, are to be applied.

ISBN 0-9519236-2-5

British Library in Cataloguing Publication Data.
A catalogue record of this book is available from the British Library.

Published by: Milepost Research,
8 Green Bank, Barnoldswick, BB18 6XHX
01282 850430

Printed by Nayler's of Church

*The dumb boat **Mary** under tow on the Leigh branch in the 1950s. By then, boats on the canal had become very run down, but the wooden ones continued to be painted in the traditional style. (Edward Paget-Tomlinson)*

Contents

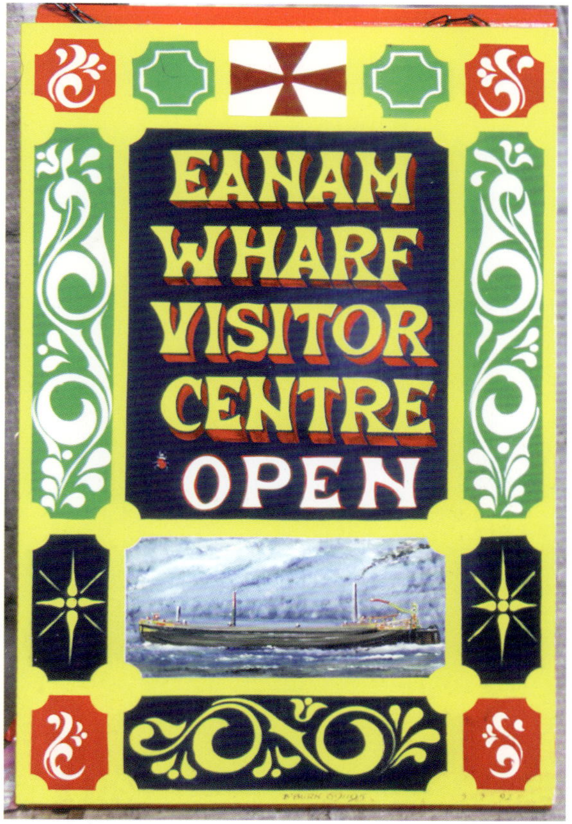

A sign painted by Sam Yates in 1992 in the style used at Whitebirk.

What the book is all about: a fine, but comparatively simple, example of Leeds & Liverpool Canal brightwork, the canal company's ***Tiger*** *below Bingley Five-rise locks around 1895. (Boat Museum Archive)*

Introduction

The Leeds & Liverpool Canal has many claims to fame — it is Britain's longest canal and Bingley 5-rise locks are one of England's few Grade 1 listed canal structures — however, despite this, the canal's unique history and culture has never been widely recognised. It never closed, so has never had a restoration programme, with all the publicity that requires, and there is generally a lack of awareness of the cultural and economic importance of Lancashire and Yorkshire. Perhaps this is why I, an industrial rather than a cultural or art historian, am writing about the decoration of Leeds & Liverpool Canal boats. Hopefully this book will encourage those with more knowledge of the history of the arts to continue the research.

My interest in the canal developed in the early 1970s when I bought a traditional wooden Leeds & Liverpool Canal boat. For five years I lived in the boatmen's cabin, the last surviving example still with its traditional grained paintwork, whilst attempting to restore the boat to something like its original condition as built in 1939. I was already aware of the roses and castles of the narrow canals of the Midlands, having worked with Tony Lewery, the expert on the subject, so realised that the decoration on Leeds & Liverpool Canal boats was completely different. However, the energy required for maintaining a large rotten wooden boat left little time for research into its decoration. It was only years later, after writing the history of the canal, that I made a contact who made possible research into the decoration of boats on the canal.

In 1989, a rally of former Leeds & Liverpool Canal working boats was held at Eanam Wharf, Blackburn, in association with the small canal visitor centre which used to be based there. Amongst the crowd of local people, I was fortunate to meet Sam Yates who had begun his working life at Hodson's boatyard at Whitebirk, building and repairing the boats which carried coal between Burnley, Wigan and Blackburn until the hard winter of 1963/4. Fortunately, old Mr Hodson had passed on to Sam the skills of painting the traditional decoration used on the boats, and he was able to describe these, as well as producing detailed drawings of the designs, identifying the colours used for the different panels and shapes. As with many craftsmen, he was surprised that anyone else was interested in the details of his work, but was delighted to be able to pass on his knowledge.

Over the following years, we have often met to discuss the canal's traditions. I once traced a boatbuilder in Yorkshire who had also painted boats on that end of the canal, but he died before he could pass on much of his knowledge. Fortunately, this was not the case with Sam, who has been able to record his experiences at Hodson's. In 1999 we wrote a history of the yard and its workings, published in ***Aspects of Blackburn***. Now we continue with a book which reviews the origins of the traditional boat painting on the canal, its variations, and similar painting elsewhere in Europe. The style and methods used at Whitebirk are discussed in detail, hopefully with enough information for those who would like to preserve the craft or to use it on boats today. The style of painting at Whitebirk is also compared to that used elsewhere on the canal, particularly in West Lancashire.

Many people have helped during the research for this book, but I would particularly like to mention Tony Lewery and Edward Paget-Tomlinson, who encouraged me to look into the subject in the first place. Tony has provided much additional information and subsequently commented on the results of the research.

Illustrations and suggestions have come from many sources, with many of the former from my own collection. Others come from The Boat Museum Archive, Edward Paget-Tomlinson's collection, Tony Lewery, Harry Arnold, Harry Belshaw, Phil Speight, Peter Worden and Philip Watkinson. It is unfortunate that few photographs were taken specifically of the decorative paintwork, and I have had to rely upon enlargements of small details, making some of the images in the book a little blurred.

Mike Clarke
Barnoldswick, 2009

*The **Pluto** in 1975. The name panel and scrolls were painted at Wells' yard in Stretford, with the rest of the decoration undertaken by Mike Clarke when just beginning his research into the canal's history and traditions.*

*The **James** of Crook & Thompson, immediately after launching in 1947. Note that the registered number had not yet been issued to the new boat. The photo comes from a cine film of the launch, now available on DVD. (Peter Worden)*

***George** in the livery of the Wigan Coal & Iron Company after restoration at the Boat Museum in the late 1980s. Although the colour scheme is comparatively simple, the deck furniture — barrel, proven box, ventilator and kennel — are all well decorated. The colour scheme was extensively researched by Phil Speight and is considered to be authentic. The orange red, almost vermillion, was thought to be the 'best red' for use on boats at yards in the Wigan area. A well-covering red was needed as the top coat could suffer if lower colours 'bled' through, a continual problem even today.*

With all the boatman's family members aboard, the ***Plato*** *is seen here at Burscough. John Parke's business was based in Liverpool, and the* ***Plato*** *shows many typical decorative features for the area, from the very stylised lettering to the picture of a house on the registration panel. Was this related to the boat being registered to allow people to live aboard as it features in other photos of boats from Liverpool? Elsewhere on the transom are many scrolls on either side of the registration panel, with candy stripes on the middle guard iron. As would be expected from boat decoration of this quality, there is a picture in the panel on the rudder head. (Philip Watkinson)*

Traditional Paintwork on Leeds & Liverpool Canal Boats

Today the Leeds & Liverpool Canal is used just for pleasure, with most activity on the canal taking place in the summer. Narrow beam pleasure boats are the main traffic, and increasingly they have little or none of the decorative paintwork often associated with canal boats. Such decoration requires regular attention for it to maintain its attractiveness, work which today's boat owners often consider excessive on a boat dedicated to leisure. For some owners, there was a similar attitude when broad-beam cargo-carrying boats were the mainstay of the canal, with a wide variety in the amount of decoration. But all boats had some decoration, however simple.

In winter, with rain dashing down from grey Pennine skies, life on board Leeds & Liverpool Canal boats must have been hard as the boats glided up and down the locks carrying essential cargoes for smoky industrial towns like Wigan, Blackburn, Burnley, Bradford or Shipley. Even the terminal towns of Leeds and Liverpool, despite their fine architecture and philanthropic pretensions, were depressing places for manual workers. Fortunately for those on the canal, the countryside between Lancashire and Yorkshire's industrial centres is some of England's finest. Could it have been this which gave boatmen on this canal a special appreciation of form and colour, for some of their boats were perhaps the best decorated of any, whether canal or sea-going. Arguably they were superior to those of their colleagues on England's narrow canals, as well as those from the Tagus estuary or the salty coastal lagoons of Portugal, where even today the sun shines down on boats with brightly coloured

*The **John** of J & W Tomlinson, Blackburn coal merchants, after unloading at Whittle. The decoration, with extensive use of scrolls on the transom, is typical of that found in East Lancashire, and unsurprisingly is halfway between the Yorkshire style and that of the Liverpool to Wigan area. (R. Thomas)*

Brightwork was a tradition mainly found on wooden boats, and today few boats on the canal are painted in the style. The iron and steel boats always had less decoration, but on the ***Pauline****, seen here at Goole, the tradition is being revived.*

decoration. Yet, despite the fact that the last of the painters of 'Brightwork', the name given to decoration on the Leeds & Liverpool Canal's traditional boats, only ceased work in the early 1960s, thirty years later the tradition had almost been forgotten. Hopefully this book will encourage people to resurrect and value this attractive style of boat traditional decoration.

On the Leeds & Liverpool Canal, boat painting was typified by geometric patterns, intricate borders and edgings, scrollwork, and colourful scenes or floral decoration. The finish was of high quality even though it was just a small part of the work of the one or two of the boat-builders at a boat yard who maintained the tradition. Often it was only the yard owner himself who did the detailed painting and sign writing, whilst any of his skilled workers could undertake the basic ground work. The designs were handed down from generation to generation, with each boat yard having its own recognisable style within the boundaries of the tradition.

After the First World War, the amount of detailed paintwork seems to have declined, particularly when iron and steel motor boats were introduced. At the same time, competition from road transport and the decline of traditional canalside industries led to a decrease in the profitability of canal transport, with boat owners probably cutting costs by reducing the amount of paintwork. However, a couple of boatyards continued painting boats with traditional decoration into the early 1960s, only finishing when traffics on the canal's main line ceased following the hard winter of 1963-4.

Wherever the tradition came from, the colourful decoration of English canal boats is one of their most striking features. Until a few years ago, the traditional styles used on the Leeds & Liverpool Canal were only recorded by old photographs and their significance was recognised by just a few old canal workers and canal historians. Better known to the general public was the canal boat painting tradition of 'Roses and Castles' which developed on the narrow-boats of the Midlands. The design of narrow boats is not connected with maritime traditions, and this is reflected in the decoration which also has no connection to maritime traditions.

However, there is much more to canal boat painting, and Edward Paget-Tomlinson's ***Colours of the Cut*** gives a much more comprehensive overview and indicates that, for narrow boats, roses and castles were not universal. This was especially so in north-west England, where narrow-boats on the Shropshire Union, Trent & Mersey and Bridgewater canals had a different style of painting to that associated with narrow-boats in the Midlands. Even there, the day boats on the BCN had their own style, which used geometric patterns quite extensively. On boats everywhere in the early twentieth century, the standard of painting was high, narrow-boatman Charlie Atkins telling the author of the excellent finish to the painting of SUCCo boats prior to 1921. The simplistic paintwork now often associated with narrow boats, only developed after the World Wars when maintenance costs had to be reduced. As a result, the amount of paintwork undertaken by the dock yards was also reduced, with boatmen themselves adding their own naïve style of the decoration. Roses and castles could be found on some boats but, scrollwork, geometric patterns, graining and simple bold colours were also a feature of north-west narrow-boats, and these seem to reflect the boat painting of the coastal boats around the Mersey Estuary. They certainly bear more than a passing similarity to the painting on Leeds & Liverpool Canal boats, where all those features, plus several more, could be found.

*The **Irene** of the Shropshire Union Canal has grained doors and a carved name. External decoration is simple, but done to a high standard. Some roses can be found, but few compared with the narrow boats of the Midlands, (Boat Museum Archive)*

A Unique Tradition

As it links Lancashire with Yorkshire, the design of boats on the Leeds & Liverpool Canal was influenced by two traditions. Construction of a canal from Leeds to Liverpool began at either end in 1770, and it was not until 1816 that the two sections met. During those forty-six years, the boats using the two sections of canal, one in Lancashire and the other Yorkshire, had developed somewhat differently, and these differences, both in

*The **Eliza** of Tarleton, whose transom is a fine example of the West Lancashire style of decoration, with scenes filling the central panels on each side. They would also have contained the registration details. Scrolls are used around the elliptical registration panel as well as in other panels on the stern. On deck, the cabin ventilator and the dog kennel would also have been decorated using traditional patterns.*

the way they were built and their decoration, were still apparent almost 200 years later. By then some Yorkshire conventions had been adopted in Lancashire, and vice versa, so over the two centuries, the canal has formed a route not just for cargo, but also for skills and traditions.

In Yorkshire, boat design developed from keels, the sailing craft of the Humber Estuary. In fact, Leeds & Liverpool Canal boats originally worked down the Humber to Hull, so they would have been fitted with masts. These were probably removable and stored somewhere on the Aire & Calder Navigation when the boat was working on the canal. Keels were originally clinker-built, each plank overlapping its neighbour and clenched to it by spikes which had their ends riveted over. Carvel-built keels and canal boats, where the planks were spiked to frames and the plank-edges butted up against each other, were only widely introduced in Yorkshire towards the end of the nineteenth century. Clinker construction was found to be a problem on canal boats, as the riveted spike heads wore away, rubbing away on the stonework whilst boats passed though locks. Yorkshire-built boats were recognisable by their very full lines, with very rounded shapes to the bow and stern which maximised carrying capacity. The bluffer lines of the Yorkshire keels may reflect their use of lee-boards, which helped reduce leeway when sailing empty. Lee-boards were a feature of many coastal sailing craft which worked around the North Sea.

Boats on the canal in Lancashire developed from Mersey flats. From early in the eighteenth century, these were carvel-

Left: Boats on the Lancaster Canal were more likely to be made of iron and had much less decoration than those of the Leeds & Liverpool Canal, despite being somewhat similar in design. Perhaps the lack of decoration was a reflection of the relative prosperity of the two canals, though the Lancaster was also a little more isolated from maritime influence.

Below: An Aire & Calder Navigation dumb boat at Stanley Ferry showing the simple use of undecorated panels and carved stern rails typical of coastal sailing vessels around Yorkshire. However, the rope fenders over the stern rail suggest that the paintwork was of a good standard and worth protecting.

*The **Shamrock**, iced in on the Rochdale Canal, has the minimal decoration typical of boats on that canal. The carved rudder detail is suggestive of origins in maritime tradition.*

*A square-sterned Bridgewater flat, the **Annie** of Simpson Davies & Son, passing under Monton Bridge has, like Lancaster Canal boats, little decoration on its transom beyond the boat's name, registration and owner. Simpson Davies boats did sometimes work onto the Leeds & Liverpool Canal, but were not based there.*

built, and so were boats built in Lancashire. Lee-boards were not used around the Irish Sea, so sailing craft here had to have finer 'lines', their shape under water, with the sharper bow and stern 'gripping' the water better when the boat was empty and thus reducing their tendency to slip sideways in the water when the wind was abeam. As a result, their shape was finer than those of Yorkshire-built boats. The smaller carrying capacity resulting from the more 'pointed' bow, was often offset by a square stern. This feature was in use on Yorkshire canal boats by the twentieth century, the design coming via the Leeds & Liverpool Canal.

The Leeds & Liverpool Canal was only linked to Liverpool's Docks in 1821, and even then, boats had to travel via the Leigh Branch, the Bridgewater Canal to Runcorn and then down the Mersey. A few boats undertaking this trip seem to have had removable masts, possibly left at Runcorn when travelling on the canals. The opening of the Stanley Dock branch in 1846, made such trips and masts unnecessary. Perhaps it was because the canal remained unlinked to the river for so long, that Lancashire canal boats seem to have developed their own individual style, independent of coastal vessels, to a greater extent than in Yorkshire. There the canal had joined to the Aire & Calder Navigation, and thus a route to the sea, as early as 1777. This link with existing types of coastal boat was apparent in the way canal boats were decorated, with Lancashire boats developing their own individual style, whilst in Yorkshire there were greater similarities with the decoration on boats on other canals around the Humber estuary. The style of painting may also have reflected the differences between Lancashire and Yorkshire folk. The latter revel in a dour, parsimonious outlook, possibly a reminder of the old-established sheep farming traditions of the area where money was not spent unnecessarily, and quality was essential. Lancashire folk were more innovative, adopting new methods and industries, such as cotton textiles, and there were certainly many new colours developed in East Lancashire for the textile trades. Perhaps this was a reason for the more colourful nature of decoration on Lancashire built boats, the tradition possibly reaching its peak on coal boats working between Wigan and Liverpool prior to the First World War, though boats in East Lancashire also had fine examples of brightwork.

Boatyards and Brightwork

All the decorative paintwork was undertaken by boat builders at the boatyards along the canal. Little in the way of facilities were needed to set up such a boatyard, just skilled craftsmen and a suitable piece of land alongside the canal where a pair of slipways could be laid for pulling boats from the water or for re-launching them. However, most yards were much better provided for, with covered workshops, slipways and drydocks making work easier in winter weather. Towards the end of the nineteenth century there were over one thousand boats working on the canal, and on average, each would have needed a week or so for repair each year. So at least twenty boatyards were needed to maintain the boats on the canal and, in fact, over thirty boatyard sites have been identified. There would certainly have been others.

The size of the yards varied from the basic ones employing just four or five workers to those with extensive facilities and a work force of thirty or forty. Not all boatyards built new boats, but when they did it would take four men and an apprentice around three months to complete one. With around half a mile of seam to a boat, caulking to make the hull water-tight was one of the most time-consuming jobs. Usually others helped, with up to ten people working on the boat at busy yards where they were available.

After the boat's wooden structure had been completed it was ready for painting. The basic colours could be applied by any of the yard's workers, whilst the painted graining of the cabins in Lancashire was usually undertaken by the apprentice.

In Yorkshire, varnished wood often remained a feature of the cabins until costs had to be reduced from the 1920s onwards. After the basic single-colour ground had been completed, the decorative brightwork was then undertaken. Usually there was just one man who did this work. The designs and elements used were handed down from generation to generation, with each boatyard developing its own style, instantly recognisable by those who worked on the canal. Just as the construction of the boats had developed independently at each end of the canal, so had the decorative designs. Once again, the influence of coastal craft could be seen, with carved decorative panels sometimes

*Flyboat **260** on the slip at Mayor's Tarleton yard c1910. The florid lettering is typical of the Liverpool end of the canal, and though there is no scroll work, the guard iron decoration is quite detailed.*

found on the knightheads at the top of the stem post, with the name carved into the top plank at bow and stern. Scrollwork could feature as part of the name panel. In Yorkshire, the boat's name could also appear on the stern rail above the rudder post.

In Lancashire, Leeds & Liverpool Canal boat decoration seems to have developed away from its origins to a greater extent than in Yorkshire, possibly because the canal was not linked to the river in Liverpool until 1846. Its relative isolation from coastal craft, compared to Yorkshire, may have resulted in this evolution. However, lack of illustrative material prior to the 1860s makes it difficult to quantify the particular changes and their origins.

Other factors could have influenced the designs, as some decoration was required under the canal's authorising Act of

A misty day at Whitebirk in the 1950s, with a Crooke & Thompson boat just off the slip.

1770. The name, number and owner of each boat had to be registered with the canal company and painted on the boat under section 65 of the canal's first Act (10 Geo III, Cap 114). They were to be in white capital letters, six inches high and of proportional breadth, on a black background and located on the outside of the boat above the loaded water-line. Perhaps this requirement for white lettering on a black background led to several canal companies, the Leeds & Liverpool and Shropshire Union being two examples, using black and white for the company colours of their boats. The Canal Boats Act of 1877 (40 & 41 Vict, Cap 60) further required the boat to have the place of registration and number exhibited in a conspicuous manner, the subsequent Act of 1884 requiring this to be on either side of the boat or on the stern such that it was visible from both sides of the canal. The date at which the repainting of a boat's cabin was undertaken was also often incorporated into the boat's decoration as this had to be recorded for the registration authority.

Painting was always carried out at a boatyard. All of the men could paint simple panels, whilst the sign-writing and scroll work was usually the job of one man, who, although not necessarily trained, could have many years experience. Often the painter was also the yard owner, foreman, or one of the longer serving tradesmen. These men became highly proficient, and some painted for a hobby as well. This led to an extremely high standard of decoration, of which both the painter and boatman were justifiably proud. The standard of decoration declined during the 1930s, with the introduction of iron and steel motor boats, though wooden boats built at the time maintained the old traditions to a great extent. The iron and steel boats represented a period of austerity, as it was considered that they needed less docking, their simpler painting style reflecting this. Wooden boats required more care on the part of the boatman, who were encouraged to look after them by of the high quality of the paintwork. Boatmen preferred wooden boats, which handled better, and whose cabins were more comfortable to live in, provided the deck did not leak. Wooden boats would also be repaired by the craftsmen at the canal's many boat yards, working to traditions established over the generations.

After repair, the hull was blacked using a mixture of pitch and tar, though the areas to be painted or varnished were left untouched. These areas were then scraped and filled where necessary to provide a good surface for painting, and the first coat of primer or varnish applied. Areas to be painted were now finished with the correct ground colour after suitable undercoating, to provide a smooth surface for the brightwork.

The decoration of the bow was common to all wooden boats with the top plank, above the main guard, being painted. Panels, about three feet long, on either side of the stem post, contained the boat's name, followed by one or two smaller panels decorated with scrolls, bunches of fruit, or other devices. These panels were in bold colours, sometimes grained, and surrounded by a contrasting border, with scalloped corners. The D-iron, protecting the edge of the deck, would be picked out with a variety of geometric designs, such as triangles or stripes. Photographs from early in the twentieth century show that these would sometimes have scroll work or other patterns added to them. The decoration ended as the boat approached its full width, because paintwork here would inevitably be damaged in locks and bridge holes. The

Don *was formerly steamer* ***31*** *and was motorised in 1935. Here a group of boatmen at Crabtree Lane, Burscough, pose on her bow. The style of lettering and the scrolls are typical of East Lancashire, so she may have been docked last at Hodson's yard at Whitebirk. Jack Hodson often mentioned the difficulty in painting the Liver bird, as even a slight mistake was almost impossible to rectify, and the work had to be done again from scratch. The extra line inside the panel was more typical of Yorkshire, though it was used regularly everywhere.*

top of the stem post, next to the top plank, had a single panel painted on either side, which could contain the mark of the town where the boat was registered; for example, a Liver bird represented Liverpool, a sheep for Leeds and a star for Wigan or Manchester. Underneath the main guard an area of the hull would be varnished, possibly a left over from the maritime

The size of lettering used for the boat's name reduced over the years so that it would fit onto the top plank. (Tony Lewery)

tradition. Often this was replaced by a white patch, which made the boats easier to see when working at night. On some boats the varnishing would include the plank above the rubbing strake, and it would be continued over the full length of the hull. This was occasionally done in red instead.

On a round stern, the decoration was similar to the bow, though the main guard, which did not reach to the stern, was replaced by a panel giving the registration number on the left hand side, with the town of registration on the right. The boat's name appeared only on the left hand side of the stern post, with the panel on the right hand bearing the name of the boat's owner. The stern post and the top of the rudder were more profusely decorated than the stem post, with panels containing scroll work and the like. The date of painting could also be included in one of the smaller panels.

The square transom provided the largest area for the boat painters art, which reached its peak on coal boats working to Liverpool. The boat's name was on the left side of the stern post and owner's name on the right, both painted with lettering, shadowed to the right, across the top panel of the transom. Shaded lettering on Leeds & Liverpool boats was always shadowed to the right. The area beneath this, above the lower iron guard, was divided into several panels, sometimes oval in

A horse-drawn flyboat entering the top lock at Greenberfield. Although limited in decoration, the knighthead seems to have a picture on its central panel.

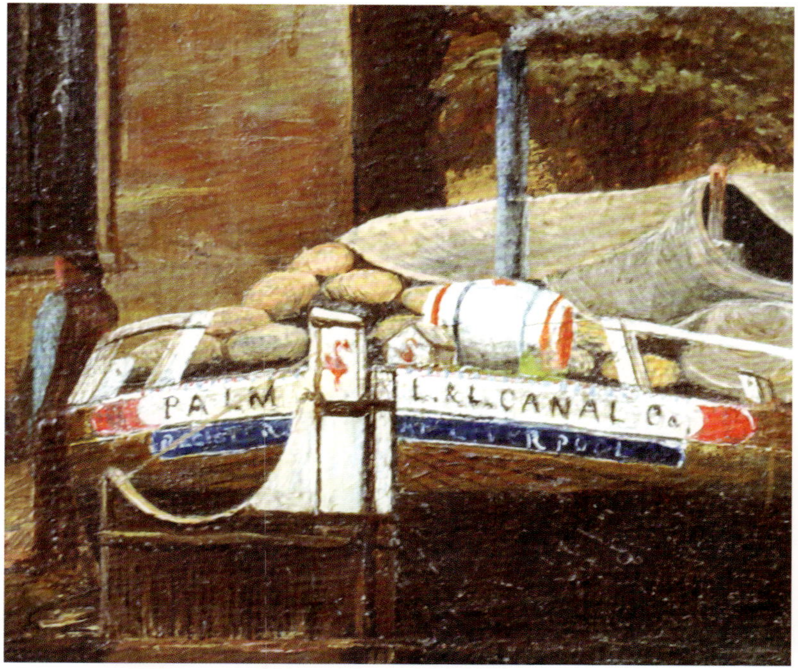

*A painting of **Palm** c1890 at Kildwick Warehouse. The red signal on the warehouse door was used to indicate that a boat needed to stop to pick up a cargo. The enlarged details give, top left, a good idea of the company's colour scheme, and bottom, the carved detail and the use of natural wood finishes on Yorkshire boats.*

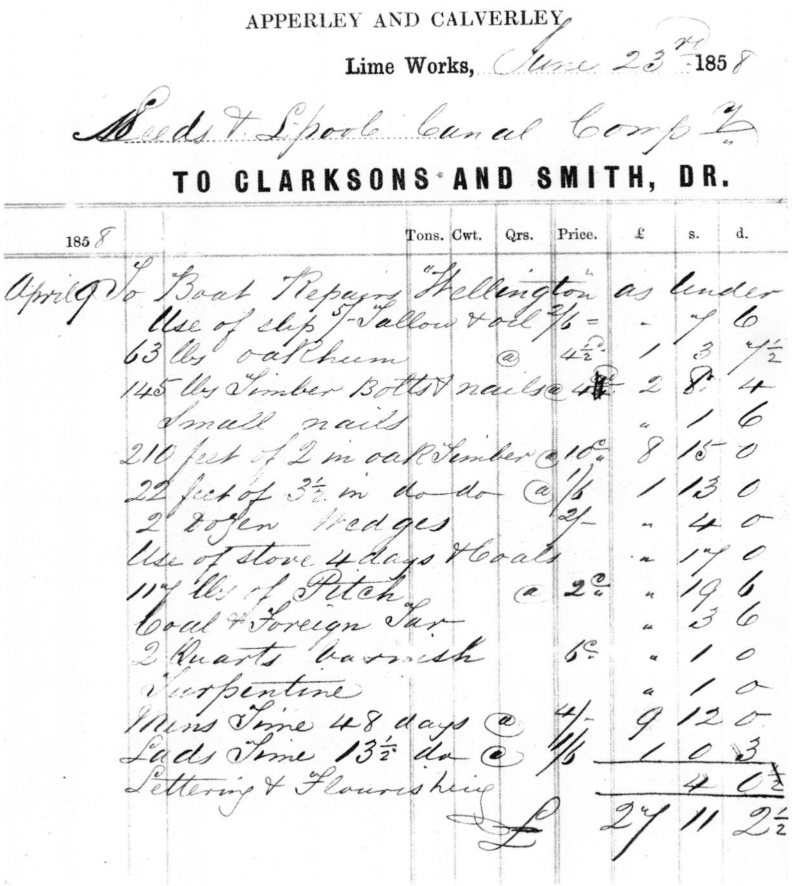

APPERLEY AND CALVERLEY

Lime Works, June 23rd 1858

Leeds & Lpool Canal Compy

TO CLARKSONS AND SMITH, DR.

1858		Tons.	Cwt.	Qrs.	Price.	£	s.	d.
April 9	To Boat Repairs "Wellington" as under							
	Use of slip 5/- Tallow & oil 2/6 =					-	7	6
	63 lbs oakhum @				4½	1	3	7½
	145 lbs Timber Bolts & nails @				4d	2	8	4
	Small nails					"	1	6
	210 feet of 2 in oak Timber @				10d	8	15	0
	22 feet of 3½ in do do @				1/6	1	13	0
	2 dozen Wedges				2/-	"	4	0
	Use of stove 4 days & Coals					"	17	0
	117 lbs of Pitch @				2d	"	19	6
	Coal & Foreign Tar					"	3	6
	2 Quarts varnish				6d	"	1	0
	Turpentine					"	1	0
	Mens Time 48 days @				4/-	9	12	0
	Lads Time 13½ do @				1/6	1	0	3
	Lettering & Flourishing						4	0
					£	27	11	2½

A bill for work done in 1858 at Apperley Bridge on a L&LCCo boat. The pay for a boatbuilder was 4/- per day, and that is the amount allowed for the 'lettering and flourishes' itemised on the bottom line. Out of 48 days work, only one was allowed for decorative paintwork, though work days were longer then.

***Nidd**, built at Yarwood's in 1934, and **Comet**, built at Wigan in 1938, at Skipton. Iron and steel boats had much simpler decoration, with fewer scrolls than on wooden boats. (EWPT collection)*

*The **Ben** of William Knowles, simple paintwork, with just the registration details in the main transom panels, and a Liver bird in the smaller side panels.*

Flyboat **253** *at the top of Bingley Five-rise c1910. She was built at Salterforth in 1905 and has scrolls and lettering typical of East Lancashire. Note the slight serif to the main lettering and the more italicised registration lettering. The colours used on canal company boats were basically black and white, with red, blue and green only used for detailing and on the coamings, chimney, covering boards and deck.*

shape. The number of panels could reach seven or eight on each side on the most lavishly painted boats. One panel would contain the registration details, again possibly including the registration authority's emblem, though this could appear in a separate panel. The other panels often contained the usual scrolled patterns, though on some boats there would be a picture in one of the panels. Besides bunches of flowers or fruit, there could be still life portraits, animals, or landscapes; as one retired painter said: 'You'd a job to think of something new'. Somewhere, usually at the top of the stern post, the date when the boat was painted would be included, as health inspectors could be quite strict in following the regulations governing the regular painting of cabins.

The decks, fore and aft, were painted with red lead, while the covering boards, around the edge of the deck, and the cabin hatch covers were green or blue. The side decks were often white, as this increased their visibility at night, though the toe boards, at the edge, were usually red. The bollards were also red, with white panels to make them stand out at night. The decoration would be completed by a curved design, often in blue, on the deck behind the stem or stern post. Coamings, where fitted, would be painted in a strong, contrasting colour, such as blue or green. On motor boats, the cabin sides could contain the owners name, but they were never decorated to the same extent as the transom on horse-drawn boats. Even less decoration appeared on the iron and steel motor boats, possibly because their paintwork was far more vulnerable, as their smaller iron guards would not protect the paintwork as much as those on wooden boats.

As well as the iron guards to protect wooden edges, rope fenders were used to protect paintwork. Often two sets were available, a working set made from old rope, which inevitably became worn, and a set used for show, made from cotton and scrubbed white. The latter would not be used when the boat was moving, because of the risk of them becoming blemished. Protective decoration also took the form of flat wooden shapes six or eight inches across, using designs such as hearts or spades, sometimes edged with cotton rope, and painted with scrolls or other patterns. Some boatmen did not like to use fenders whilst moving, considering they were the sign of a man who could not control his boat. They were certainly not used early on in the canal's history, as the committee complained, in 1815, about the damage caused to locks, by iron work fitted as protection at bow and stern. Such ironwork was ordered to be removed, and boatmen told to use rope straps to stop their boat from touching the gates while passing through locks.

Finally, the items on deck, such as the water barrel, dog kennel, cabin ventilator and proven box, would be painted.

***Martha Alice** was operated by Ainscough's of Burscough, one of several millers to have fleets on the canal. The decoration is quite simple, except for that on the guard irons across the square stern below the name. The panel lining is simple compared with that found in Yorkshire, where corner scrolls were often used. The registration is in simple central panels on either side of the stern post, possibly with scrollwork in the outermost panels. The only other decoration is a scene at the top of the rudder post, which has been capped by a cotton rope turks head, with further rope decoration down to the stern of the rudder. Ainscough's were proud of their boats, so the paintwork would be of high standard, though the style is a little plain.*

These were sometimes the property of the boatman, who had them decorated at the boatyard. Similar designs to those on the transom were used, with the addition of the boatman's initials in the centre of a panel. Much was often made of the decoration of the water barrel, with the ends receiving some of the painter's best work.

***George**, shortly after restoration at the Boat Museum, Ellesmere Port. The deck furniture — barrel, proven box and dog kennel — have all been stored on a plank across the hold forward of the rear head ledge to facilitate cleaning the deck. **Scorpio** to the right, has also just been repainted by Barry Thompson, who worked with Harry Leyland, a boat painter from Parbold. (Tony Lewery)*

A fine example of 'brightwork', the ***Richard*** *of Liverpool coal merchants Richard Williams. With that name, was she the owner's pride and joy? The decoration certainly suggests this as it is full of detail used on boats from the Liverpool end of the canal. The Liver bird is an obvious item, and the two houses on the registration panel feature in other photos of well-painted West Lancashire boats, such as Ainscough's* ***Leo****. A flower has been used on one of the outer panels, with an unusual scroll in the outermost. There are both simple and mirrored scrolls, while the guard irons have geometric patterns as well as a long scroll. No wonder the boatmen were willing to pose for a photo.*

*The **Annie** after launching from Thomas Whitaker's yard at Apperley Bridge in 1911. She was the last boat built for Robert Foster, a coal carrier based in Shipley, as his firm ceased trading after the First World War. However, back in 1911 he must have been especially proud of his new boat as the standard of decoration is probably of the highest for a Yorkshire-built boat. The work is exceptional, with each panel full of small detail. The scroll work is very different to that found in Lancashire and is much more bound up in the lining out of the panels. A variety of colours have been used, but I suspect that some areas are grained or varnished as a raw timber finish was much more common in Yorkshire. As on Lancashire boats, there is a scene in the top panel of the rudder, with a geometric pattern on the end of the barrel.*

Lancashire and Yorkshire: the styles compared.

*Another view of the **Annie** just before launching at Apperley Bridge, 1911, with the varnished sections of the hull visible on the right edge of the photo. Note how the panels are not uniform in shape, and it was rare for two to be decorated identically, even when they were similar in shape or a mirror image. The board across the rear of the hold has been left undecorated, probably because it was likely to become damaged when the boat was loaded with coal.*

A Yorkshire boat loading stone near Salterforth. No scrolls can be seen, but the knighthead and wooden chimneys both have the well lined panels typical of the Yorkshire end of the canal. Wooden chimneys were a feature of Yorkshire boats, an undecorated band allowing for the necessary overlap when placed on top of each other.

The top of Bingley Five-rise during frosty weather, with Silsden Co-op's ***Progress*** *and* ***Industry*** *caught in the ice, with Sugden's* ***Roy*** *behind. The decoration of all three boats is typical of Yorkshire boatyards, with simple panels well lined out. The barrel on* ***Progress*** *has a simple picture, possibly of a house and stream, a common subject for painters along the whole canal.*

Barron's ***Mary*** *is also a typical example of the Yorkshire style, the cotton rope decoration used to protect the paintwork suggesting that this was of good quality. The board at the rear of the hold which allowed coal to be piled up higher than the deck is also lined, as are the stern rails, while the barrel has geometric patterns of a Yorkshire style.*

The ***Ellen*** *of Wigan Coal & Iron Company waiting to ascend Johnsons Hillock Locks has simple decoration with just a scroll in panels on the outer ends of the company name, with the star representing its Wigan registration further out. The company's boats were not lavishly painted, but there is some decoration on the upper guard iron, as well as on the kennel. (Boat Museum Archive)*

The ***Ivy****, a former carrying boat, is seen here in 1908 at Skipton when in use by the maintenance department. Decoration has been reduced, but the lettering and scrolls on the left are of the style found in East Lancashire, perhaps from the canal company dockyard at Finsley Gate, Burnley.*

Mary Alice at Barrowford, possibly going for a load of stone from Salterforth. The decoration is simple but effective, with several scrolls and well lined out panels, typical for East Lancashire. Decorative ropework has also been made to protect the paintwork.

The Mary of Wheelton, seen here at the top of Johnsons Hillock Locks, was registered in Leigh. The decoration, with numerous scrolls, is typical of East Lancashire. There are pictures, in the small panels either side of the transom and at the top of the rudder head, but these are not as arresting as the scrollwork. Note that the guard irons have no decoration, and neither does the proven box. Is that a small dog kennel to the right of the bending man, or is it a large cabin ventilator? (Peter Worden)

The ***Ann*** *was a Tarleton flat, probably owned by Suttons, and has particularly fine scrollwork around the stem post. The second panel down appears to have a picture of two people, presumably one of them being Ann. The name panel is slightly unusual as it is formed by scrollwork, though this has been found in other photos. The paintwork was probably done at Mayor's yard at Tarleton.*

The ***Four Sisters*** *was owned by Suttons of Tarleton around 1910 and used for carrying stone from Parbold quarries to the Ribble where it was used for training walls. She is a flat rather than a canal boat, but has been decorated using the L&LC style. The scrollwork around the knighthead, upper stem post and name panel are good examples of those found on the Liverpool end of the canal, as is the lettering. Maintenance work on the boat was undertaken by Mayors of Tarleton, so the decoration is probably by them.*

Appleby's ***Leopard*** *at Blackburn, with paintwork probably from a dockyard between Wigan and Liverpool judging by the style of scroll and the flowery lettering. Note the extensive scrollwork on the rudder and the scene on the rudder head. Just visible on the barrel are the captain's initials, together with some scrollwork.*

Ainscough's ***Leo****, a 72 foot square stern boat here in the 1920s near Scarisbrick. It is a pity we cannot see more of the decoration as it is a particularly fine example of the style found around Liverpool. The registration surrounds a cottage scene which appears on other boats from this area. The name has the florid lettering typical of the area, with a fine detailed design on the guard below the firm's name. Flowers can be found in other panels, with a Liver bird in the outermost panel. The deck vent also has a flower motif. (E. Paget-Tomlinson collection)*

__Wm Robinson__ at the top of Bingley Five-rise c1895 is typical of the company's steamers which had less decoration than the dumb boats, especially at the stern where paintwork could be damaged by the boats being towed.

__Catterick__, built by Mayor's at Wigan in 1949, seen here during delivery trials. Perhaps because it was closer to the office where costs were controlled, the decoration has been much reduced. Some of the scrolls have been simplified and some panels, such as those on the stern post, have no scrollwork. The name on the stern plank has been carved and painted with no shading, as was also the case with the next boat to be built, __Darlington__, the last of the wooden general cargo boats and built at Wigan in 1951.

*Hargreave's **Frank** around 1970 when wooden boats were finally being removed from service. By then the paintwork had become much simplified, the scrolls being very thin and the geometric patterns just a ghost of their former glory. However, even at this late stage, the tradition had been maintained, though the painter was not boatyard trained as the letter shading is on the 'wrong', lefthand, side of the letters. (Tony Lewery)*

*Steamers rarely had much decoration, and the **Walter** of John Robert Thornton from Skipton was no exception. No scrollwork is visible, with just simple panels and lettering on the engine room.*

*A contemporary model of the steamer **Ruth** from the collection at the National Waterways Museum, Gloucester, shows the simple paintwork of steam boats.*

Maintenance Boats

The square sterned maintenance boats often had a single large scroll on each side of the transom as seen here on the ***Annie*** *and the* ***Mary****. Photos show the former at its base in Skipton in the 1950s and the latter at Finsley Gate, Burnley, c1940. The second photo of Annie was taken c1930 when Anchor Bridge at Gargrave was being replaced. Patterns were sometimes used for such scrolls which is why they can be quite similar. At Whitebirk there was a simple pattern made by painting over an old scroll with red lead and then applying thin card to the paint. The resulting pattern was not used to produce an exact copy, but as a visual guide when setting out a new scroll.*

Icebreaker **22** at Riley Green. The type of lettering used for the number was typical for Leeds & Liverpool Canal Company boats.

Tugs, such as the tunnel tug **Foulridge**, *had little decoration, though the lettering was of high quality.*

After nationalization, the surviving general cargo boats were painted in British Waterways' blue and yellow colour scheme. The iron and steel boats transfered to the maintenance fleet continued to be painted in this simple style, often with a transfer roundel such as is seen here on **Severn** *near Chorley around 1970. (John Gavin)*

*The **Frank** at Johnsons Hillock in 1958 after a recent repainting. The boatmen were the Turners, well-known for looking after their boat with care, and they certainly would have appreciated the continuation of the traditional painting, complete with its scrollwork, geometric patterns and varnished timber, unique to the Leeds & Liverpool Canal.*

The Origins of Scrollwork

The scroll was central to the decoration of Leeds & Liverpool Canal boats, and for generations it has been an important feature of traditional boat decoration elsewhere. Maori war canoes have circular work and scrolls on their bows, while the Phoenicians favoured horses heads on their ships' bows. The Egyptians had eyes painted on either side of a boat's bows, the boat's curved ends taking the form of a lotus stem surmounted by a leaf or flower. Eyes feature on boats elsewhere in the world; possibly they hoped they would be a benefit to their fortunes in navigation, a mixture of skill and luck for early mariners. In the nineteenth century, European and American sailing ships often had decoration called gingerbread work, sometimes associated with a figurehead. The long flowing scrollwork could well have come from the acanthus leaf, found in several countries around the Mediterranean. Why it was used is open to question, but it was an important feature on the boats around the Atlantic. The scrollwork still used on Thames sailing barges was of this type, and their method of construction, with adjacent planks rebated and tarred to form a water tight join, is indicative of an origin in southern Europe. Why have such decoration? One reason seems to have been to ensure good luck for the boat and those on board.

And what about scrollwork which is found not just on boats, but also on buildings across Europe. Perhaps it is to 'Celtic' traditions that we should turn, for a more 'local' source for scroll-based decoration in Europe. Outlines of circles can be found on some Celtic crosses, each overlapping circle representing aspects of religious life and society. The Anglo-Saxons also used a type of scrollwork on their crosses and other religious monuments, so here too is another possible link to a religious meaning to the decoration. Scrolls may be a stylised form of these groups of circles, which provided some form of religious or cultural meaning. Then over the years they lost their religious significance and became simply decoration.

Simple circle-based designs, scrolls and decorative edgings, like the 'dragons teeth' used on D-iron protection to the covering boards forming the edge of the deck of Leeds & Liverpool Canal boats, can be found right across northern Europe, not just on boats, but also on buildings. Examples still decorate houses as far away as the Kargopol region of northern Russia, an area under the

Circles of influence which were representative of society and religion in 'Celtic' times, with suggestions as to how these could be translated into scrollwork. Such symbols could have been regarded as a form of protection for the boat and its occupants, and this may be the reason that scrollwork has been passed down as a feature of boat and house decoration.

influence of the Orthodox Church. Decoration is an important feature of Orthodox churches, wall paintings and icons being set out in panels, in a way comparable to decoration on the square transom stern of some Leeds & Liverpool Canal boats. The Catholic Church also used iconography and decoration as part of their religion, something that was repressed by the Protestant religion in the north west of Europe.

If boat decoration does have religious origins, we would expect to discover similar decoration to that found on Leeds

Scrolls and geometric patterns feature in the painted houses of Northern Russia such as these examples from the Kargopol region. The simple carved scrolls around the windows can still be purchased as mass-produced items, though their possible origin as religious symbols has been forgotten. Even the interior of the houses were decorated, and the tradition seems to extend across the north of Europe. In the Friesian area of northern Holland and Germany.

The Orthodox Church in Russia used panels extensively in their decoration of their churches. Perhaps this association played a part in the development of decorated panels on boats, or the origins may have simply been from the way objects were made.

& Liverpool Canal boats elsewhere. Probably the only other European country where such extravagant boat decoration has survived is Portugal, where the moliceiros, which carried salt, and sea weed for fertiliser on the Ria around Aveiro, and the fregatas and botes, the sailing barges of the Tagus around Lisbon, have a long tradition of painted decoration.

The fregatas carried passengers and general cargoes across the Tagus at Lisbon, forming an essential link between the capital and the industrial areas to the south of the river. This traffic virtually ceased in the 1960s following the opening of the first bridge across the Tagus in Lisbon, and the traditional paintwork almost disappeared as a result, with just one painter still at work in the years after the bridge opened. Fortunately a few boats survived, and today more are being restored, preserving both the boats and their decoration. Several painters are now at work. Their overall style of decoration is similar to that found on the Leeds & Liverpool Canal, with specific panels and planking used

Boats on Russian rivers sometimes used decoration similar to scrollwork, such as on the stern of this boat photographed about one hundred years ago.

Boats across Europe used scrolls for decoration: this is a dghajsa from Malta.

*Leaves and vines emulated scroll work, or was it the other way around? Above, the bow of the **Jane Banks**.*

*Left, are two Scottish fishing boats, the lower with a more defined religious symbol included, and at the bottom the Thames sailing barge, **Hydrogen**. (Tony Lewery)*

Travellers also used scrollwork, perhaps for the same reasons — religious protection as well as decoration. (Tony Lewery)

The moliceros of central Portugal featured panels with religious and cultural scenes. This Baptism of Christ, now preserved in the Maritime Museum at Ilhavo, was copied from a boat in the 1930s.

Today it is still possible to find religious scenes on the bow or stern of moliceros, but with the advent of mass tourism you are more likely to find paintings of subjects similar to the saucy McGill postcards of seaside Britain.

Religious symbols, such as the Madonna here high up on the bow overlooking the working area, are a regular feature of the beach-launched fishing boats of the Portugese coast. Such religious icons have survived in Catholic areas, whereas in the Protestant north of Europe they have been replaced by more secular images.

for the boat's name, registration, owner and for decoration. This suggests that in both countries boat owners were keen to identify their boats and to use the decoration almost as advertising as well as providing something of which the boatmen would be proud. Similarities also exist in the variation in the amount of decoration; in both countries some firms spent much more on decoration than others. On the Portugese boats, the borders to the panels often have scalloped corners, just as on the Leeds & Liverpool Canal, and although the flowers and other decoration are different, it is just as detailed as that shown in old photos of the best examples of Leeds & Liverpool Canal decoration. Larger panels include images of religious significance, local scenes or flowers.

The moliceiros have a different style which is also set out to a specific pattern, with panels for identifying the boat builder/painter, the owner, place of origin, and for decoration. There are four main panels, either side of the bow and stern, and originally these displayed religious images or local scenes and personalities. More recently, with the increasing importance of tourism, panels with cartoons, similar to the seaside postcards of Donald McGill, are the most common, and one of these usually wins the annual prize now awarded for the best decorative panel. Around the panels, the borders reflect the use of the boats, with designs related to plants and water. The modern paintwork reflects the changes in culture, the influence of the Catholic Church declining as Portugal changed from a closed dictatorship to a democracy. The rise of tourism then brought different influences to the decoration.

Both types of boat had religious influences on their decoration, and this would have developed from a need for

Other designs, such as flowers, simple geometric patterns and edgings, are used to decorate the Portugese boats.

The fregatas of the Tagus estuary are perhaps the closest to Leeds & Liverpool boats in terms of decoration. They still have some religious panels, but there are also ones for the registration and ownership of the boat as in England. Scalloped edges to the panels are often found, as are fancy borders, though these have evolved in a different way.

Decoration at bow and stern are the most important areas, with the panels filled in much the same way as in England, though with different designs and colours.

religious protection. Even today, traditional fishing boats in the area have a carved Madonna looking down on the boat from the bow, and old photographs of fregatas often show religious images. Fregatas almost certainly used carved decoration originally, as found on many sea-going vessels. Painted decoration seems to have been introduced from the mid-nineteenth century. Did painted decoration develop as paint technology improved? The boats in Portugal, as on the Leeds & Liverpool Canal, were based on sea-going vessels. However, their use on inland or sheltered waters allowed paintwork to survive much longer, and without degradation, compared to those vessels sailing the saltier sea. As they were tied-up regularly to tranship cargoes or people, it would also have been much easier to touch-up any blemishes, with many more opportunities for this compared to sea-going vessels.

Can we see similar influences affecting the traditional paintwork of Leeds & Liverpool Canal boats? For centuries, the major religious influence in England has been the Protestant Church, in which colourful religious decoration is frowned upon. Possibly as a result, the paintwork on boats developed a less religious style compared to Portugal, with secular subjects being used to replace religious ones, as the basis for decoration. The changes which have taken place recently in the style of boat painting in Portugal could have happened one or more centuries ago on the inland waterway boats of northern England. Religious motifs, which had been designed to fit the blank spaces at the bow or stern of such boats, were altered and adapted to suit the more secular style which developed on Leeds & Liverpool Canal boats.

One problem in deciding the origins of Leeds & Liverpool Canal boat painting, is that there are no images of boat decoration before the advent of photography in the late nineteenth century. Certainly one source for brightwork is marine decoration, which probably had some religious significance.

Not all boats on the canal had elaborate decoration, with some just using simple scrollwork and rectangular panels. This may have been the style used on the earliest canal boats and survived in a simpler fashion on other northern canals. The more detailed designs, recorded on late-nineteenth and early-twentieth century photographs of Leeds & Liverpool Canal boats, could have been a new development. In the mid-nineteenth century, Augustus W Pugin was beginning his revival of Gothic-style architecture. In places, he used scrollwork and geometric designs, which bear some similarities with those on Leeds & Liverpool Canal boats. He designed Scarisbrick Hall in 1836, close to the canal in West Lancashire and not far from Burscough, where many of the canal's boatmen lived. This Gothic revival was taken up everywhere, and the roof of the church at Addingham, not far from the canal at Silsden, has a good example of this school of decoration, painted in the second half of the nineteenth century. Perhaps an increasingly detailed style of Leeds & Liverpool Canal boat painting was part of this revival, the designs being influenced by the ecclesiastical style which Pugin favoured. The subsequent Arts and Crafts school of William Morris, where mundane everyday objects were profusely decorated, could also have encouraged owners to adopt elaborate paintwork for canal boats.

The rapid improvement in the brightness of paint pigments and dyes during the second half of the nineteenth century could also have influenced boat decoration, especially as much of the research took place in East Lancashire.

The decoration on the inside of Addingham Church roof dates from the late 19th century. Could this revival of Gothic patterns have influenced the decoration on Leeds & Liverpool Canal boats? Has a simple traditional pattern been enhanced by painters and boat owners influenced by the then current trends?

Decoration on the Leeds & Liverpool Canal

Leeds & Liverpool Canal boats, developed from the coastal craft of the Mersey and Humber, were influenced by local maritime traditions. As they sailed on fresh water, canal boats provided a better place for traditional decoration, provided it was protected from damage in locks and at wharfs. From the owner's viewpoint, decoration would not only appease the boatmen's sense of tradition, but would also encourage them to look after the boat carefully.

The decoration probably came from designs with religious significance to help protect the boat and crew. These became less overt because of the influence of Protestant ethics. Then canal Acts required specific panels to be used for the boat's name, registration and owner. This developed into a traditional style, with panels being used for specific types of decoration as well as information about the boat. The finest examples may date from the decades around 1900 when trade on the canal was at its height. It could have been influenced by the rise in the 'Arts and Crafts' movement. Only a few boats and owners seem to have used the full possibilities of the tradition, with a favourite boat sometimes receiving such attention. The majority of boats had simpler designs, but most still retaining some form of the traditional designs.

To sum up the variations amongst owners and along the canal, the amount of detail used in the paintwork would depend upon the individual boat owner and how much they were prepared to pay. The time of year could also influence this, painting being much harder in the winter and time more difficult to justify when the boatyard was busy. Some owners

***Crucis** is seen here circa 1955 near Foulridge, steerer Tom Cheetham, with James Draper watching. Scrollwork is limited, with none on the stern post or the rear of the engine room. The lettering of Canal Transport Ltd has not been laid out evenly, not uncommon when there was little time for paintwork. The mirrored scroll in the panel next to the company name is typical of the Liverpool end of the canal.*

*The **Lizzie of Shipley** on the slip at Apperley Bridge, near Bradford, around 1910. The transom is painted in the typical Yorkshire style, with simple panels surrounded by extensive lining. The boatbuilder, G. E. Ramsey, subsequently moved to Shipley.*

would have one of their boats extensively decorated while similar boats in the same fleet were plainer, while other owners would just have simple decoration. There was great variation in the extent to which square transoms were decorated, and the designs and styles also varied along the canal. Scrolls also varied from yard to yard, with some using fine lines best seen from close up, whilst others had bolder strokes which made the scroll easier to appreciate from a distance.

In Yorkshire, the decoration used graining and lined panels with intricate corner designs to a much greater extent than in Lancashire. The boats were probably less colourful, though the standard of painting was always extremely high. Varnished woodwork could also feature, grained panels rather than varnish probably representing a decline in the quality of the woodwork. On Yorkshire keels, boat cabins remained varnished long after this tradition had disappeared on smaller canal craft. Along the whole canal, items such as the water barrel were decorated with scenes or geometric patterns on their end panels, but in Yorkshire the curved surfaces seem to have been less brightly coloured.

On the Liverpool end of the canal, geometric patterns, decorative scenes and brightly coloured panels predominated, the whole being developed into a unique style to a greater extent than in Yorkshire where the style still had definite links to that found on coastal boats. Between Liverpool and Wigan, scrolls would include mirrored examples, the design being repeated around a vertical or horizontal axis. Lettering done at boat yards in that area was often extremely flowery, with curled serifs in the middle of a letter as well as at the end of each stroke.

Church Wharf c1910, with a canal company steamer and dumb boat ***111*** *moored inside. Decoration on canal company boats was never particularly extensive, but there were always sufficient scrolls to keep within the tradition. Panels on steamers rarely had scalloped corners which gave them a more utilitarian look. Both water barrels have scenes on their end panels, with scrolls on the centre of the curved sides. The square stern boat passing was registered at Blackburn, and its decoration, with many scrolls filling the panels, is typical of East Lancashire boatyards at this time.*

As would be expected, in East Lancashire there was a mixture of the two styles. The colouring was probably brighter than in Yorkshire, but with fewer of the scenes found around Liverpool, with geometric patterns and scrolls predominating. However, it was in East Lancashire that the customs of boat painting on the Leeds & Liverpool Canal survived longest, with the boatyards at Riley Green and Whitebirk continuing to use traditional decoration, albeit to a somewhat reduced extent, right up to the end of regular cargo carrying on the canal's main line in 1963. By then Parke's fleet was the only one carrying coal to Liverpool, and little painting was being done at their dockyard at Bankhall. Similarly, the boats carrying coal to Wigan power station from the Leigh branch had little paintwork, the majority being steel boats formerly used for carrying grain.

Paintwork at T & J Hodson's

It was only the small East Lancashire fleets of Hargreaves, Crooke & Thompson, and Dean Waddington which continued to be decorated up until the end of their operation. They were built, repaired and painted at the yard of T & J Hodson, at Whitebirk, near Blackburn, or at Crook's of Riley Green. Sam Yates, a boat builder who was apprenticed and worked at Hodson's, remembers how the painting was undertaken during his time there.

The basic painting was done by all and sundry, whereas the signwriting was mainly done by the owner of the yard. Each apprentice was allowed one attempt at putting a yellow line around the engine cabin. If it was a reasonable sort of a line, then he could do the top stroke. This could go on for maybe two or three years, and if you were good enough then you would be introduced to lettering and so on. Fortunately I had a good eye.

I took over the signwriting proper when Jack Hodson had a bad spell when he was aged 72. He passed his brushes on to me and said, 'It's all yours now'. He did not check on my work, it was not the done thing to say whether it was good or bad, it was just a grunt of approval or disapproval. Although, as an apprentice you were taught the skills of boat-building, Mr Hodson felt that signwriting was something '**in**' a person.

With regard to lettering, I just 'got on with it', if it looked right, it was right! Although my work was very similar to Mr Hodson's, I had my own recognisable style. We did have some guides for lettering, very much like a stencil, for the names of the firms — such as Crook & Thompson or Dean Waddington — as this could save time. The shape of the plank to be signwritten was cut from card and the lettering sketched out. Holes were pin-pricked through the card at the edge of the letters. Then the card was held against the plank and a bag of chalk tapped over the stencil. This left dots of chalk on the plank, and these could be used as a guide for painting. The card would perhaps be given a coat of shellac to keep it stiff, and then hung up for use on the next boat from that firm. It just saved the time of setting out the letters.

Few transoms were decorated in my time, but I did find out how to decorate barrels, provender boxes and such like. There

Jack Hodson, the yard's owner from the 1930s. He undertook all the detailed decorative paintwork until handing it over to Sam Yates. Note the deck decoration and scalloped edged panels on the square-sterned boat, ***D'Arcy Wyvill****, which has just been rebuilt.*

Whitebirk boat yard, and Sam Yates poses at the tiller of a Dean Waddington boat, with Jack Hodson behind the engine room. Today the open country in the background is covered by industrial and retail units.

New boat ***Alice*** *on the slip at Hodson's around 1890 just before launching. James Hodson is stood by the stern and Jack Hodson (cousin to the later yard owner) is holding a maul on his shoulder. The yard also did work for the colliery in the background and other local firms.*

were no set scenes for barrel-ends, etc, although in the paint cupboard a few faded Christmas cards, perhaps sixty or seventy years old, and one was of a windmill, an appealing picture and another I recall of a snow-covered cottage. Quite often a painting would be something out of one's head. Occasionally a boatman might ask for something different, and I remember a boatman asking for a scroll on the rudder iron. It would only have taken a minute, but the boss said, 'Out of the question, everyone will want one'.

The cabins were grained, but had no other decoration. The oak graining was always combed, while the mahogany was brush grained. There were mahogany deck beams and carlings and shoulder pieces, and the locker tops (or seat) were pea green, with red lead underneath, and an inch band on the floor around the edge of the supports and walls, the rest of the floor being left unpainted to be scrubbed white later by the boatmen. The pea green was also used for the beading of the panel work, separating the mahogany graining from the light oak graining.

Before painting the cabin on an old boat it could be stoved to kill off any bugs, lighting the fire and putting a sod over the chimney and adding a handful of sulphur to the fire and then closing the scuttle. Afterwards the cabin was limewashed, a handful of lime to a gallon of water, and left to dry overnight as this killed any grease. You would paint straight over the lime. If we were limewashing a workshop, a very rare happening, some cow dung was added so the wash would not brush off on your clothes. For a brilliant white, dolly blue would be added.

The paintwork made the boats look attractive, and boat owners probably hoped that their men would be encouraged to look after the boats, besides keeping the boatmen happy. This was just on wooden boats, the iron boats had a lot less decoration.

The paint used for signwriting was stainers, and we mixed all our own paint. Cookson's white lead was the foundation, then there was red lead powder, turps or white spirit, sun-driers, boiled or raw linseed oil. For signwriting we always used genuine turps, and for the colours, stainers in paste form mixed with linseed oil, turps and sun driers. The emerald green was a poison and had to be signed for in the register before use. Alstons supplied us with their own brand of stainers, but now they just supply factory produced paint.

An owl lived in the boat shed above one end of the slip, and its droppings sometimes resulted in problems with the paintwork.

Hodson's yard circa 1958, probably during Wakes Week when the boats working on the local coal traffics could be docked. One has just been launched after a refit and repaint.

For signwriting there were some small tins of lemon chrome and permanent red, which we called 'bull' red. When painting bollards red we would use an orange, red lead, undercoat and then red stainers. But before the last coat of varnish we would paint it with 'slick', 70% varnish and 30% red stainers. With the orange undercoat it made a very bright red, a feature of our yard.

A boat ready for launching at Hodson's in the 1950s. This photocopied photo gives some idea of the conditions under which the decorative paintwork was undertaken, though no photo can repeat the freezing conditions of a hard winter's day at the yard.

The boss was keen on his white, so we would always add a hint of blue to make it stand out. This countered the yellowing effect of the varnish. Transparent varnish was used on the 'brightwork', the signwriting on stem and stern. The name probably originated from the varnished areas, where the planks were scraped and varnished at stem and stern. The planks there were caulked and the nails punched in and then payed with pitch. Afterwards the pitch was scraped off clean and the planking varnished. Preparation for painting was done with scrapers made from old files in the smithy. No sand paper or masking tape was used. Wire brushes and chipping hammers were used for metal plates, with a caulking iron often pressed into service as a putty knife for filling holes.

Two coats of paint were used on the decks, the first coat scrubbed well in using well-worn brushes. All the best red surfaces and engine room cabin sides had one coat of varnish. For the brightwork areas there would be two coats of paint and one of varnish, two coats on scraped timber. After the first coat of paint, the surface would be puttied-up prior to the second finishing coat. Late summer mornings were a difficult time for getting a good finish because of the willow herb drifting on the wind, while varnishing tended to be done later on, just before finishing, as there was less dust in the air. If an area had been skimped during painting or varnishing, it was called 'a holiday', time off or wasted being frowned upon.

Boat names and dates of docking were often added to the list on the inside of the boat shed .

The brushes used for signwriting would include a 'rigger' for the yellow line, and two or three sable brushes — a 'duck' (no. 6), a 'goose' (no. 8) and a 'swan' (no. 10). The names originated from when bristles were placed into a quill of the said bird — a 'lark' being minute, and a 'swan' quite large. The writing brushes, when not in use, were kept in shape using a little slip grease, and were thoroughly cleaned before reuse.

For the greater part of the year, particularly during the balmy days of July and August, signwriting a boat was a pleasant occupation with only the ring of caulking mallets breaking the silence, the use of a radio being neither wanted nor permitted. Late afternoon, the men abandoned their caulking and would 'crew' in the workshop, there hand spinning oakum for the following day. With the others in the workshop, I was left on my own to paint the brightwork. A quietness followed, broken only by the chirping sparrows and the drone of inquisitive wasps, attracted by the bright yellow paint and linseed oil.

Occasionally a slow heavily laden boat passed by, her captain, with tongue in cheek, would often criticise my lettering - his jibes being good-humoured and not in the least offensive.

At this time of day the old boss retired to his timber office and, in a haze of blue pipe tobacco smoke, busied himself totting-up labour and material costs for the recently docked boat. He was never 'at home' with scratch pen and paper, and

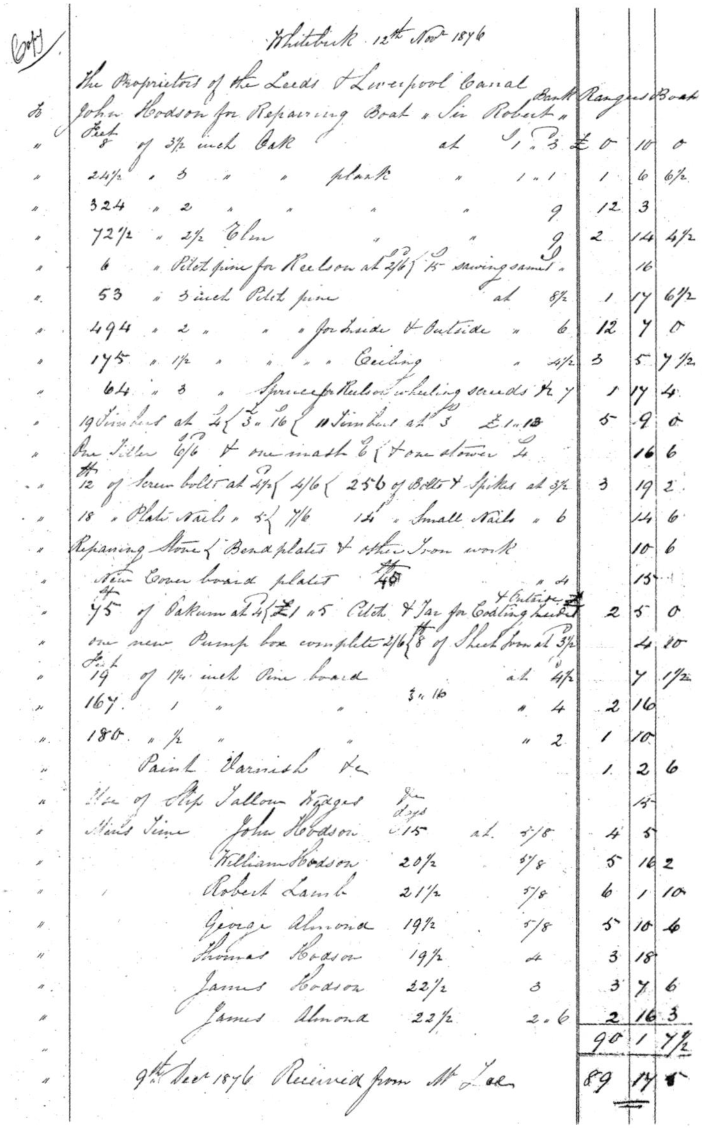
Copy
Whitebirk 12th Nov 1876
The Proprietors of the Leeds & Liverpool Canal — Bank Rangers Boat
To John Hodson for Repairing Boat "Sir Robert"
Feet 8 of 3½ inch Oak at 1/3 — £0 10 0
24½ " 3 " " plank " 1/1 — 1 6 6½
324 " 2 " " " 9 — 12 3
72½ " 2½ Elm " " 9 — 2 14 4½
6 " Pitch pine for Keelson at 2/6 15 sawing same 3 — 16
53 " 3 inch Pitch pine at 8½ — 1 17 6½
494 " 2 " " for Inside & Outside " 6 — 12 7 0
175 " 1½ " " Ceiling " 4½ — 3 5 7½
64 " 3 " Spruce for Keelson wheeling [illegible] 7 — 1 17 4
19 Timbers at 2/ 3 16; 11 Timbers at 3/ £1 13 — 5 9 0
One Tiller 6/6 & one mast 8/ & one stower 2/ — 16 6
12 of Screw bolts at 4½ 4/6; 256 of Bolts & Spikes at 3½ — 3 19 2
18 " Plate Nails " 5 7/6; 14 " Small Nails " 6 — 14 6
Repairing Stove; Bend plates & other Iron work — 10 6
New Cover board plates 40 " 4 — 15
75 of Oakum at 4/ £1 5; Pitch & Tar for Coating Inside & Outside £1 — 2 5 0
One new Pump box complete 2/6; 8 of Sheet Iron at 3½ — 4 10
Feet 19 of 1½ inch Pine board at 4½ — 7 1½
167 " 1 " " 3 16 " 4 — 2 16
180 " ½ " " " 2 — 1 10
Paint Varnish &c — 1 2 6
Use of Slip Tallow Wedges — 15
Men's Time: John Hodson 15 days at 5/8 — 4 5
William Hodson 20½ 5/8 — 5 16 2
Robert Lamb 21½ 5/8 — 6 1 10
George Almond 19½ 5/8 — 5 10 6
Thomas Hodson 19½ 4 — 3 18
James Hodson 22½ 3 — 3 7 6
James Almond 22½ 2 6 — 2 16 3
90 1 7½
9th Dec 1876 Received from Mr Lee — 89 17 [illegible]

An account of work done on a boat by Hodson's yard for the L&LCCo in 1876. No separate account was made for the time taken for decorative paintwork, just a note of the materials, showing that it was regarded as just a normal part of the work involved in docking a boat.

often found solace browsing through fly-blown 19th century ledgers — those of his late father and his father's father. At times he paid me a visit, the crackle of adze chips underfoot being my only warning of his approach. On scrutinising my efforts he never ever made comment, though his facial expression told me of his thoughts; a slightly raised eyebrow told me he

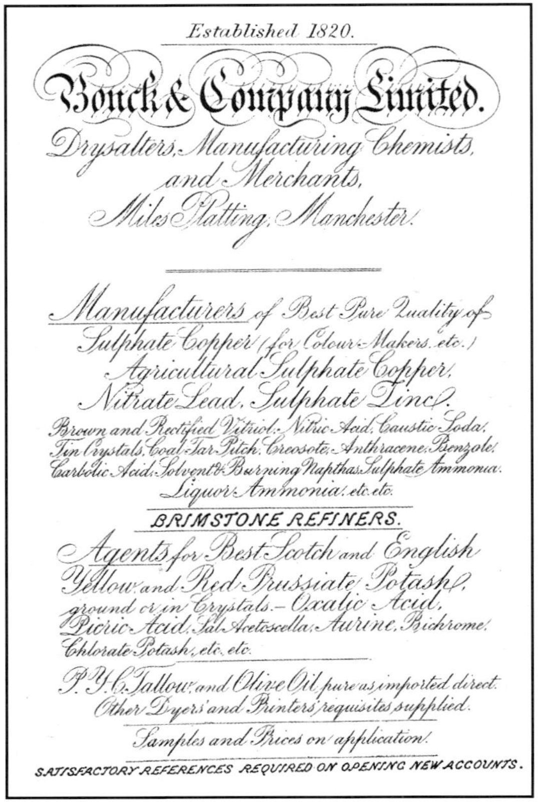

Bouck & Co were probably one of Hodson's suppliers at the end of the 19th century. Hodson's certainly repaired their boats.

***Jumbo** with Joe Bridge snr. at Whitebirk Power Station is a good example of the style of painting done by Jack Hodson at the yard.*

approved of what he saw, whereas a nibbled top lip said that something wasn't quite right. Two and a half days was allowed for completing the Brightwork, more than in old times, when the hours were longer. Roughly 22 hours were needed, but this varied with the time of year and the pressure of other work in the yard.

At day's end I walked home along the towpath, making a point in viewing my eight hours work from a distance — the criteria being simply, if it looked right, it was right. I wonder if I too raised an eyebrow in self satisfaction? I think not! After all, as the Old Man would say, 'Self praise is no recommendation.'

During the hard winters of yesteryear, signwriting out of doors became a cold and miserable way of spending several hours in combating the frost and north-east wind.

In winter it was sometimes necessary to warm the area for painting beforehand by holding a heated iron plate close to the panel requiring decoration. On a number of occasions the task proved impossible. Winter weather could affect other aspects of the yard. I recall one severe January when Maggie, the boatyard pet duck, was discovered frozen to death on her nest of frozen reeds. So too the yard's semi-wild tomcat perished in a like manner, attempting in vain to find frost-free shelter beneath snow covered logs of English oak. A far cry from times, and mice, aplenty, long summer days and warm moonlight nights.

Notes

The pallet of colours was quite restricted, with a 'bull' red, lemon chrome, a dark and light blue and a light and mid green being the only colours apart from the 'common' red as red lead was known. However the colours could vary slightly as they were made up fresh for each boat. Occasionally a different colour could be used, such as the light red and blue needed for Croasdale's boats.

*The **Jumbo**, here at Haskayne in 1972, had been painted at Whitebirk, but the decoration had been 'touched up' subsequently. However, it provides a good guide to the yard's style. (Tony Lewery)*

Brightwork at Hodson's Whitebirk Yard

__Edward__ was built in 1936 as a horse boat for the Blackburn coal merchants, Dean Waddington & Co Ltd. She was converted to a motor boat in 1945. __May__ was built in 1950 at Hodson's of Whitebirk for Crooke & Thompson, also coal merchants of Blackburn, and initially worked to Wigan Power Station, returning to East Lancashire circa 1958. In that year the two firms were purchased by the Leeds-based coal merchants Hargreaves, whose symbol in Lancashire was the flaming H, and the photo shows them in the colours of this firm. The boats were usually docked and painted at Whitebirk, in this case Jack Hodson having painted the __May__ and Sam Yates the __Edward__. (Dennis Parker)

The letterhead of T & J Hodson shows that they did not just undertake boat work, but also built sheds and repaired wagons and other structures for local industries. Millwrighting and wheelwrighting could be as much part of day-to-day work as boatbuilding and painting.

WHITEBIRK,

Blackburn, ... 193

Tel. 6041 Blackburn.

M. H. HODSON. J. E. HODSON.

M ....

To T. & J. HODSON,

BOAT BUILDERS & REPAIRERS.

PORTABLE BUILDINGS, GARAGES, &c.

John Hull & Co's colours

John Hull & Co was a coal merchant at Audley, Blackburn, with a small fleet of boats in operation from around 1900 until the 1950s.

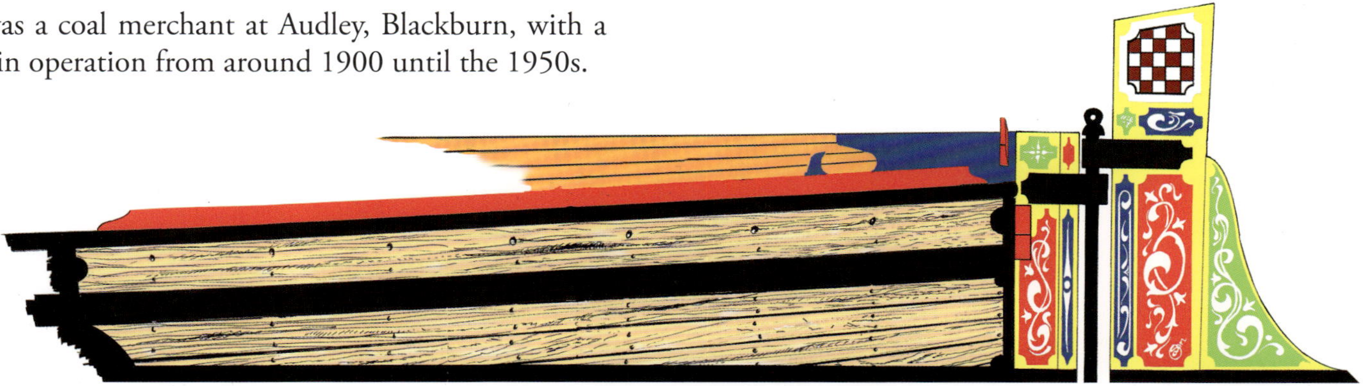

Towing masts for horse-drawn boats were fairly standard in their colour schemes, as were covering boards (green) and wash strips or toe boards (red) at the edge of the deck.

Some areas of the hull were scraped and varnished, a feature which probably came from sea-going traditions. This is illustrated here by a side view of the stern of a square-transomed boat, with the transom itself below.

John *was a square-sterned boat built in 1907, shown here without a rudder.*

Crooke & Thompson Ltd's colours

The firm dates back to circa 1825, when John Crook had a coal yard at Riley Green. In 1872 the firm amalgamated with the Thompson brothers to become Crook & Thompson, their logo, a black diamond on a white diamond, being an amalgamation of those of the two former firms. Their main yard was at Audley, Blackburn.

A typical scene on the canal, ***Joan*** *loading coal at Bank Hall Colliery in the mid-1950s. (Harry Belshaw)*

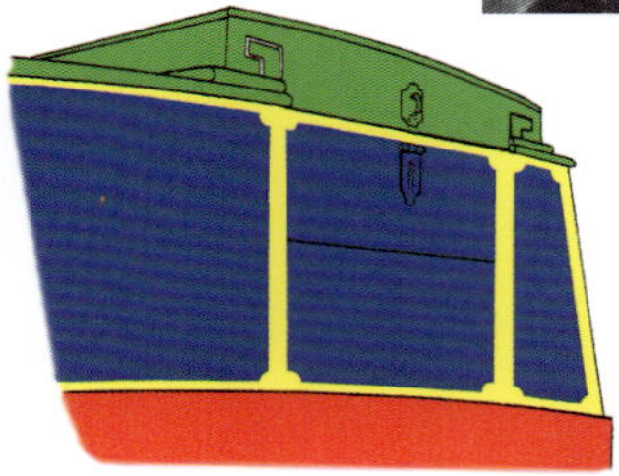

Crook & Thompson engine cabin paintwork.

Below, tillers for all companies were painted in a fairly standard colour scheme.

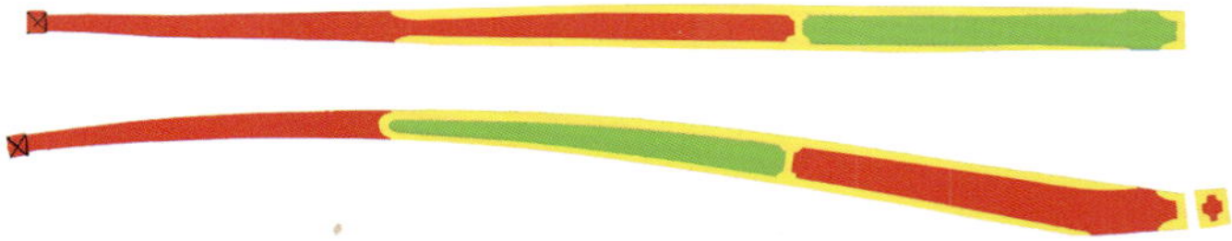

The boat's name was not painted on the bow..

Sam Yates used the Crooke & Thompson motif on the 'Leopard Award' for the L&LC Society.

Details showing the stern rail painting and scroll shading used on Crooke & Thompson boats. Note the date for when the boat was painted.

The scrolls either side of the name were shaded in the same way as the lettering, as was that in the panel beneath, the shading always to the lower right-hand side. Elsewhere there was no additional shading.

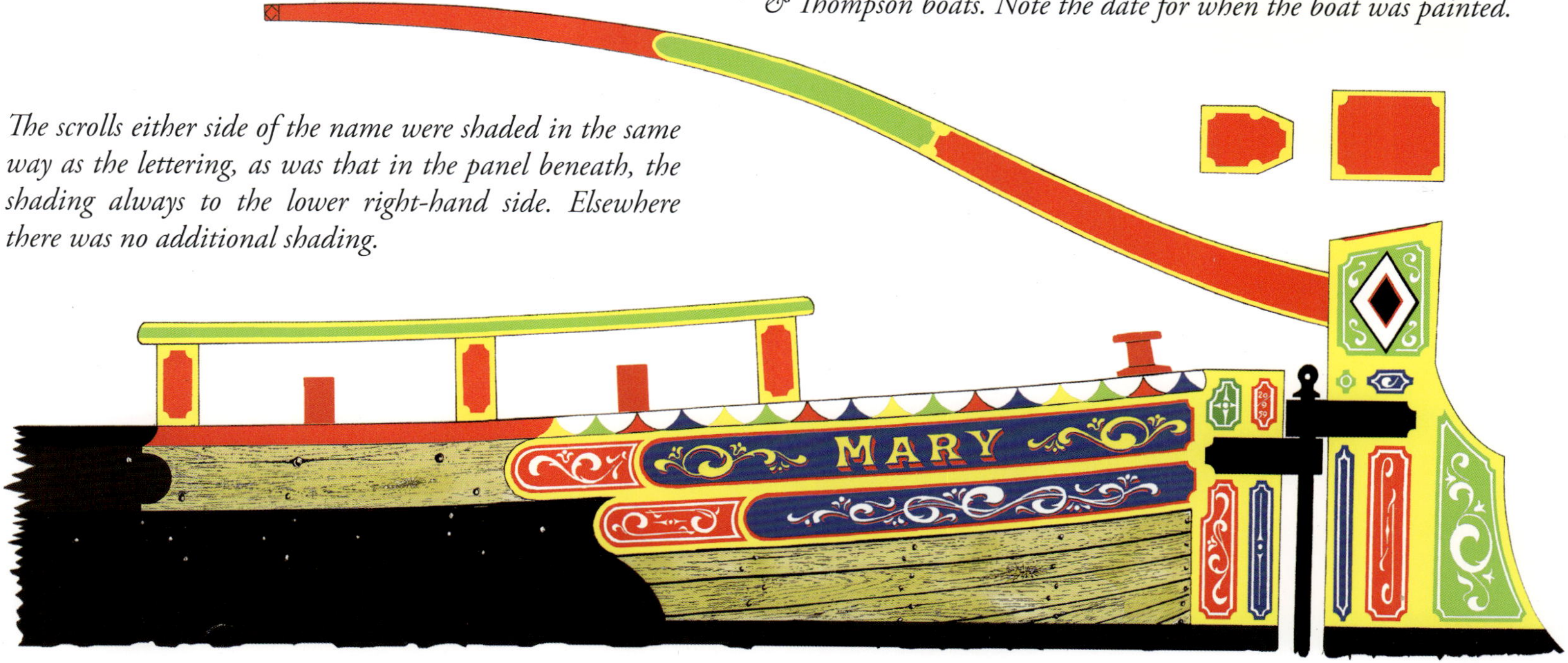

Dean Waddington Ltd's colours

The firm had depots at Eanam (Blackburn) and Feniscowles, and in the 1930s they owned 3 steamers and 15 horse boats. They worked from Wigan to the Blackburn area, with coal also carried from Burnley to Blackburn. The firm was purchased by Hargreaves of Leeds in about 1958 when their traffic up Wigan locks stopped.

Transfers were suggested for Hargreaves' flaming 'H', but at £1 each were considered too expensive. The size required also varied. The panel below the top rudder iron was often painted dark blue and left undecorated as it was subject to rust and oil stains.

*Dean Waddington's boats were also docked at Wigan. Here, the scrolls on the **Anne** are of the mirrored type so it almost certainly has been painted at Wigan yard. Note the stern deck decoration (see p.61), and the toe rails which were usually painted red.*

Henry Croasdale's colours

Based at the coalyards at Audley, the firm was in operation by 1905. When Appleby's finished carrying grain to Blackburn, Croasdale's purchased three of their iron boats, and these were less well painted, as was the norm with iron boats. The boats were taken over by Hargreaves and transferred to the Wigan area around 1960.

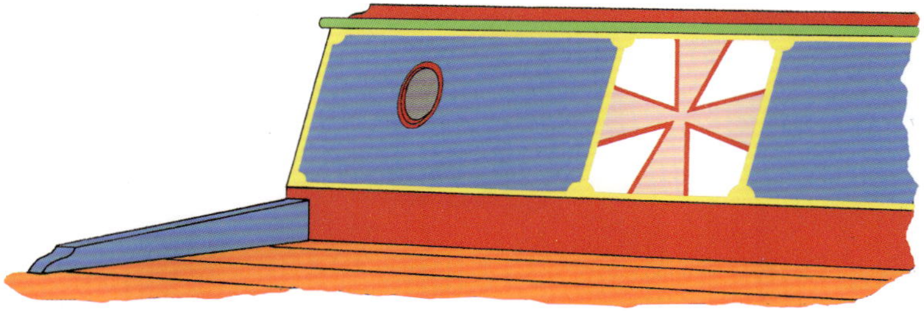

One of Croasdale's motor boats descending Wigan flight in the late 1950s. The decline in the use of decoration is very apparent, with the boat looking very rundown.

The boat's name only appeared on the stern.

Bumpers and D-irons

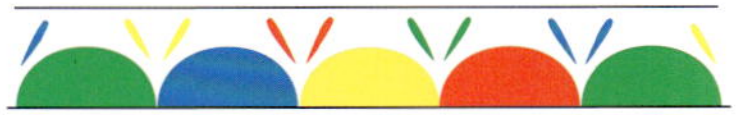

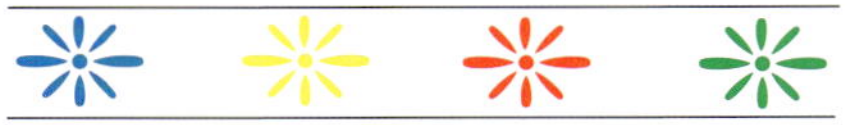

Usually, decorative patterns on the transom bumpers and D-irons were simple, but old photos suggest that on better decorated boats they could be quite intricate.

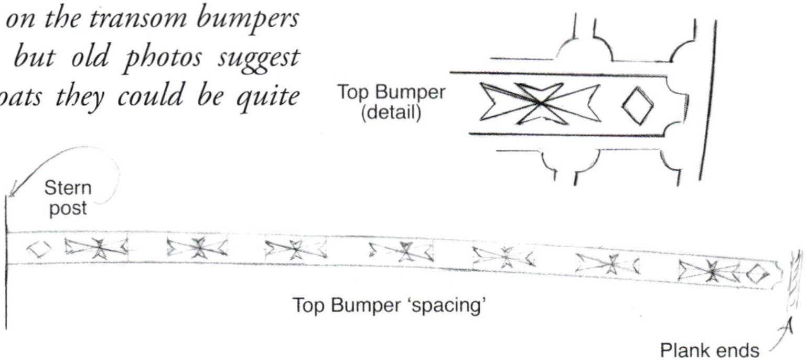

Dragon's Teeth

Scallop and Dog's Teeth

*Phil Speight, painting **George** at the Boat Museum, has used the very simple dragon's teeth design for the D-irons. (Tony Lewery)*

*Stripes, as on the **Joan** on the left, were often used in West Lancashire. Harry Leyland, whose work it is, was trained at Sheldon's Parbold Dock and, when that yard closed, moved to Wells' yard at Stretford. (Edward Paget-Tomlinson collection)*

Decks

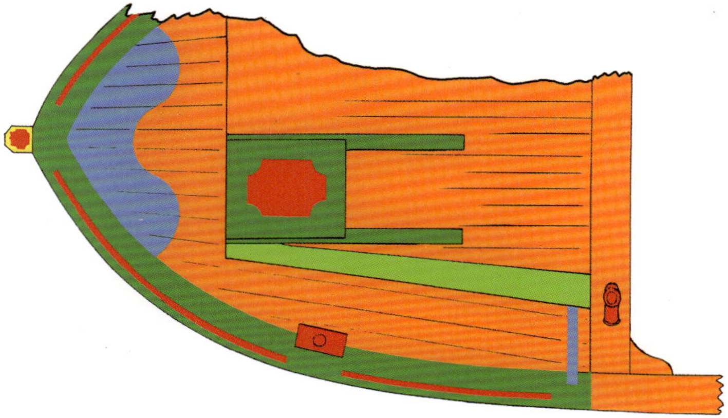

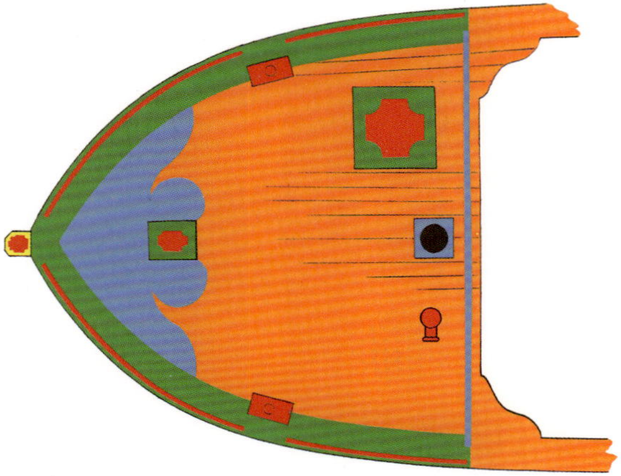

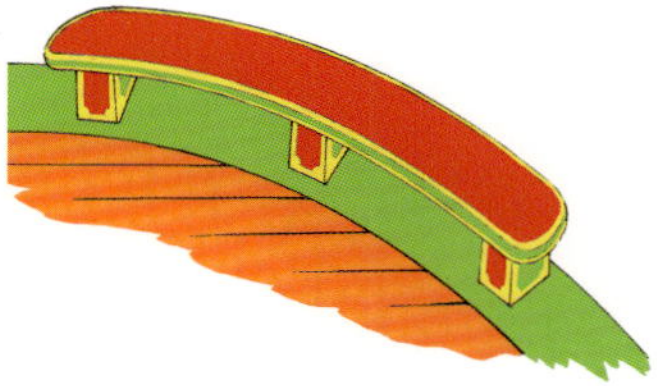

A detail showing how stern rails were painted.

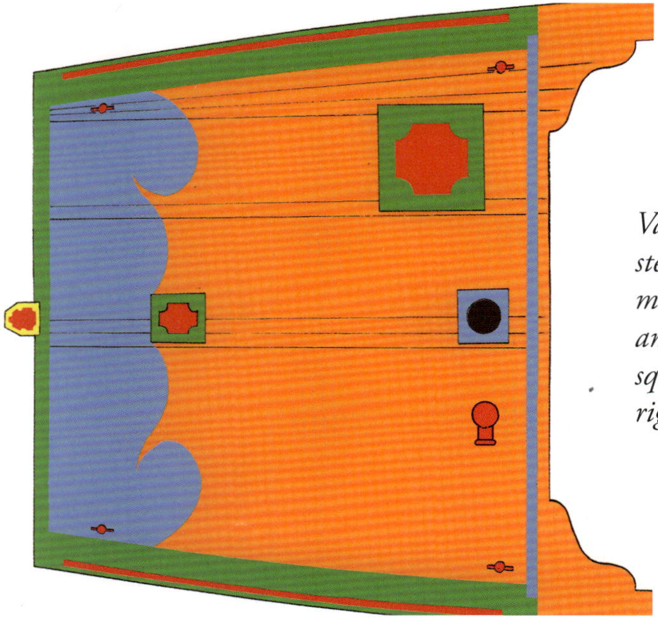

Various patterns used on the bow and stern decks. Above are the stern of a motor boat with engine room, left, and a dumb boat, right. Below is a square stern, left, and a bow deck, right.

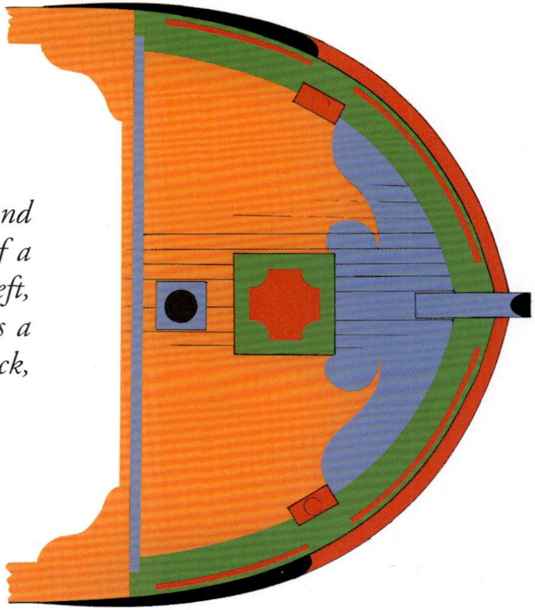

Yorkshire Chimneys

*A chimney on the **Annie** after launching at Apperley Bridge in 1911, a particularly fine example of Yorkshire painting.*

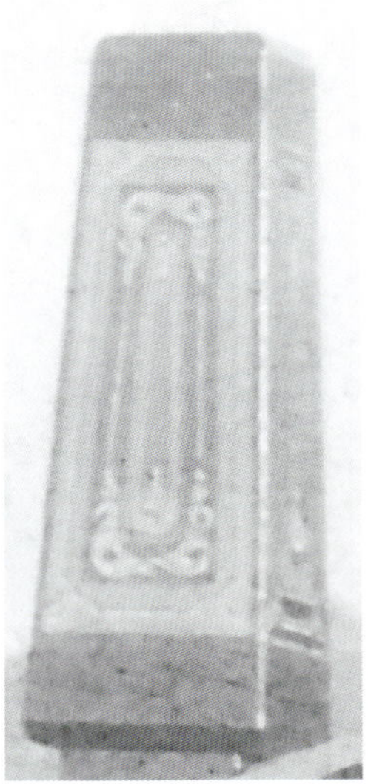

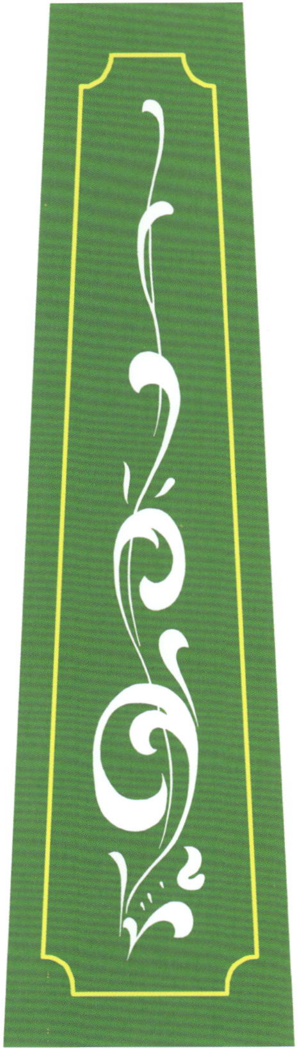

*Yorkshire-style wooden chimneys could be painted at Lancashire boat yards. The middle example comes from the **Joan**, a square-sterned wooden boat which ended up on the Bridgewater Canal. However, the chimney scroll is very suggestive of Hodson's of Whitebirk. Note that the Yorkshire painter has left room for the chimney to overlap. The drawing is from a tracing of the chimney's other face by Tony Lewery.*

*William Oldfield's **Industry** at Salterforth around 1900. Note that the chimney panelling does not extend to the top of the chimney as this allows for the overlap of the upper chimney when in position.*

Cabins

Above: Sam Yates' drawing of a single-sided cabin with double bedhole for family boating. Below: Tony Lewery's drawing shows how the table cupboard worked.

Above is a mid-1970s photo of ***Pluto****, the last short boat cabin to retain its original paintwork, though by then with much blistering. It had probably been painted at Wigan in the 1950s, and the colour of the graining is reversed to that used at Whitebirk. (Tony Lewery)*

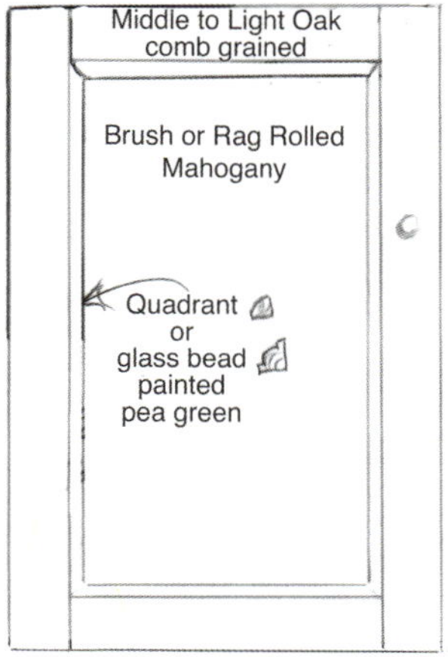

On the left are details of how a cabin was painted at Whitebirk with light oak and mahogany graining.

Water Barrels

Boats featured on barrel ends painted at Whitebirk for many years, as this 1890s photo shows.

This barrel was painted by Sam Yates for the National Waterways Museum at Gloucester.

The designs on the left are all by Sam Yates and follow the East Lancashire style. The subject often came from old postcards or Christmas cards, and a small collection was kept at the boatyard.

A barrel painted by Harry Leyland with scrolls in his Parbold style. The end panel features a house, a regular subject on both barrels and transoms in West Lancashire. The individual panels here were traced by Tony Lewery and have been coloured digitally.

Scroll by Harry Leyland of Parbold
on a L&LC water barrel stand.
Boat Museum no. C150/2

Above and right: Billy Wells had this barrel painted, though he admitted it is a little over-embellished. The scrolls are typical of the Wigan area, while the flowers on the end panel could have been painted at a boat yard anywhere on the canal.

Scrolls: Whitebirk

Three examples of scrollwork and lettering on boats repaired at Hodson's in the 1950s, the ***Mary*** *by Sam Yates and the* ***Edward*** *by Jack Hodson. The style of scrolls and the lettering can readily be recognised as coming from the yard. These illustrations are from photocopies: it is a pity the Post Office lost the originals!*

Marjorie *descending Johnsons Hillock locks in the 1950s. She was one of Crooke & Thompson's boats, as can be seen from the black and white diamond motif on the bow. The scrolls around this are unusual as they were Sam Yate's first solo effort at decoration. His later work was more in keeping with the tradition!*

*Dean Waddington's **Jenny** after repair at Whitebirk around 1960. She was a former general cargo boat which had been motorised in 1948.*

Some freehand scrolls of the style used at Whitebirk by Sam Yates.

Scrolls: Parbold and Stretford

A typical 'mirrored' scroll, often painted by boat yards between Wigan and Liverpool, here recreated by Edward Paget-Tomlinson.

*The **Jeanette** on Wells' dock at Stretford. The lettering and mirrored scrolls are typical of boat yards in the Wigan to Liverpool area. (Harry Arnold)*

*Right: a scroll copied from a photo of Albert Blundell's **Edith**, which had been painted at Stretford. The leaf motif was used quite regularly as a way of filling space. Did it come from the acanthus leaf?*

Barry Thompson, above, worked under Harry Leyland at Stretford and picked up the style developed by Harry when he worked at Sheldon's dock, Parbold. The scrolls are of the mirrored type painted at boat yards between Wigan and Liverpool. (Tony Lewery)

*This scroll, copied from the **Roland** in 1972, is too thin and was probably painted by a vehicle sign-writer.*

A selection of scrolls drawn by Barry Thompson. Note the mirrored scrolls and the more extensive use of tails than in the Whitebirk scrolls. Based on the Parbold style of scroll, these are typical of the western end of the canal and should be compared with those drawn by Sam Yates from Whitebirk on the previous pages.

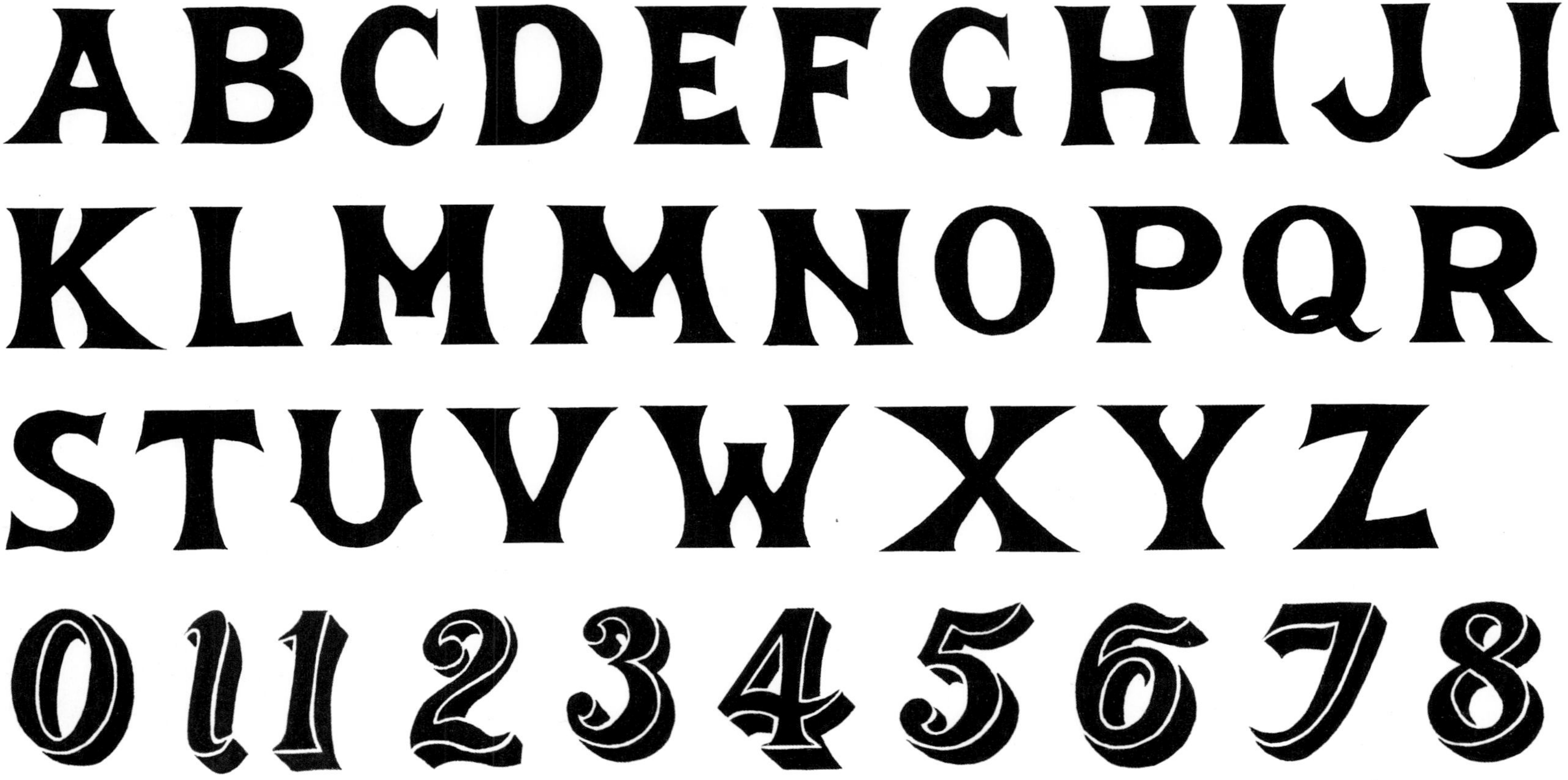

This lettering was drawn by Sam Yates and is representative of the style used at Hodson's Whitebirk yard, with a slightly curving serif to a plain face. The shading, as shown on the numbers, was always to the right lower side of the letter.

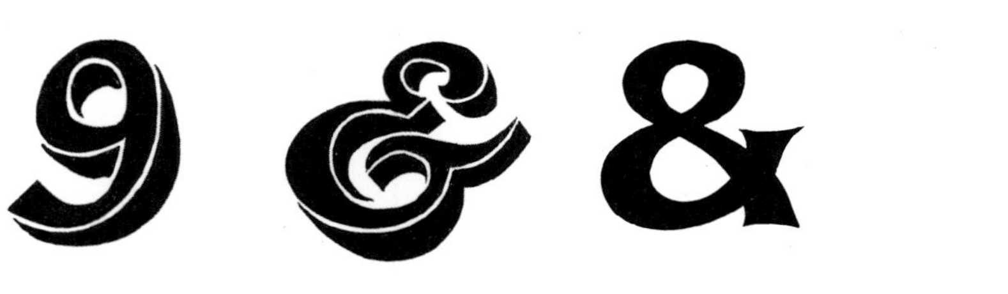

*Part of Albert Blundell's name from the **Edith**, probably signwritten by Harry Leyland.*

*The name panel of **Plato**, a Parkes' boat painted with the flowery lettering often used by boatyards between Liverpool and Wigan.*

An advert for canal transport painted for the Bridgewater Bi-centenary at Wells' yard at Stretford using very simple lettering, almost a typeface, with detailed shading.

Lettering varied from job to job, dependent on the time allowed for its completion, as shown in these various forms of 'No.' and the letter 'R' which could be used at Whitebirk.

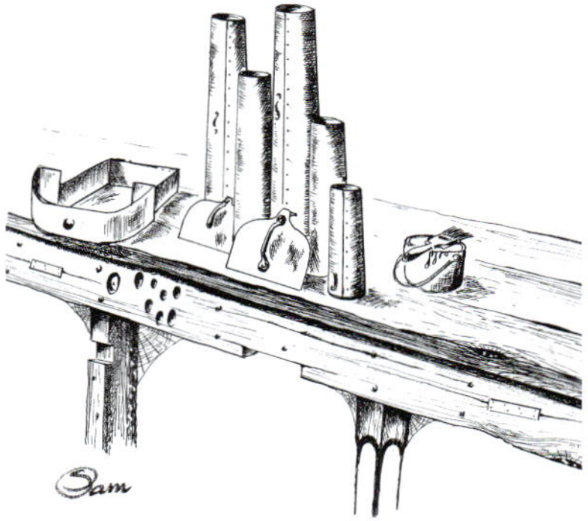

Benches were often made from parts of old boats.

Scorpio *after painting by Barry Thompson at the Boat Museum. He had worked at Wells' Stretford yard under Harry Leyland, whose style he follows. Letter type varied at Stretford. Below is a sketch Barry made for the decoration of the bow of a Leeds & Liverpool boat. It uses the mirrored style of scroll which was painted at boat yards on the western end of the canal.*

***Elsie**, one of Crooke & Thompson's boats, as painted in the late 1950s, probably by a sign writer. The shading is on the wrong side of the lettering for a genuine L&L Canal painter. The scrolls in this photo are also too thin to have been painted by someone with a full knowledge of the tradition. (Tony Lewery)*

*Below: Three photos of Albert Blundells' **Edith** after she had been bought by a group of Manchester Grammar School boys in 1969. Note the scrolls which use a group of leaves as part of the pattern, perhaps a reminder of the acanthus leaf decoration found on some seagoing boats. (Tony Lewery/Edward Paget-Tomlinson)*

Italic-style lettering used at Whitebirk.

Leeds & Liverpool Canal Boats

In 1770, when the canal was built, the size of the locks was based upon the vessels then in use. In Yorkshire, boats around 60 feet long by 14 feet beam already worked on the Aire & Calder Navigation, often working down the tidal Humber to Hull, while in Lancashire, the boats used on the Weaver, and Mersey & Irwell Navigations, were around 68 feet in length. The lock size adopted by the canal was based on the vessels used in Yorkshire, though 72 feet long boats could work between Liverpool, Wigan and Leigh.

*British Waterways took over the remaining fleet of John Parkes, and here some of the boats are moored at Wigan after the coal traffic into Liverpool had ceased. The **Ferro** has had its owner's name removed, but still has a couple of scrolls, heavily 'filled-in' centrally, typical of the Liverpool end of the canal. (Edward Paget-Tomlinson)*

Besides regional variations, there were also differences between boats for merchandise and those for mineral traffics. General cargoes were protected from the weather by covering the hold with tarpaulin sheets, stretched over three lines of supports, and wedged against the coamings around the edge of the hold to make it waterproof. For mineral traffic, no protection was required for the cargo, so they had no coamings, but a raised headledge was sometimes provided at the stern end of the hold to enable bulky cargoes to be piled up without spilling onto the aft deck. These were most common on Yorkshire boats.

Wooden boats either had a square transom or a round stern. A square transom increased carrying capacity and was often used on bulk cargo boats. There were disadvantages though: steering was more difficult as the transom could restrict the flow of water to the rudder, and they were slower through locks, as the boat had to be fully in the lock chamber before the gates could be shut. Round sterns were more common on merchandise boats, where carrying capacity was less important, and they were essential on flyboats working day and night, where speed was important. Some flyboats were only 12 or 13 feet wide, enabling them to enter locks quicker. A loaded full width boat trapped water in the lock, and with little clearance between the boat and the lock sides, slowed the boat's entry into the chamber. With a round stern, the boatmen could begin to close the gates before the boat was completely into the lock chamber, which meant a saving of several minutes passing through a flight of locks.

With so many detail variations in boat design, there were never

Church Wharf c1910 with two canal company boats unloading. Both have the mirrored style of scroll painted by boatyards on the Liverpool end of the canal, and though there appears to be little on the sides of the rudders, scrolls have been painted on the rear panels. Also the decoration to the iron guards above the name is particularly detailed, as is the end of the cabin vent on ***220*** *with its stylised rose pattern. The two boats, although superficially identical in painted style, have many differences in detail. Boatmen often had their own water barrel and it should be noted that the style and colour of the barrel decoration differs from that of the boats.*

Fly boats such as these had coamings for attaching sheets to protect the cargo, in this case bales of cotton. They were usually round sterned, though there were a few square transomed fly boats, usually slower boats working just eighteen hours a day and not non-stop. Their side decks were sometimes painted white to make them easier to see when working at night.

Panther *was built for John Hunt in 1928, probably at a Yorkshire yard which still used wooden timberheads and stern rails. Note the full bow, typical of Yorkshire boats. By the time this photo was taken in 1939 at Foulridge, Canal Transport had taken over responsibility for the boat's operation. and she had probably been docked in Lancashire. The scrollwork, with geometric patterns either side of the name and a mirrored pattern in the adjoining panel, is typical of the Liverpool end of the canal.*

hard and fast rules governing the type of boat for a particular trade. Boatmen had their own preferences, and when the canal company were rebuilding their fleet of merchandise boats at the end of the railway lease in 1873, a company director noted:

'The Shipley boats are not satisfactory, they are rather bluff and dreary, and perhaps built too cheaply. The Lancashire men seem to have a prejudice against Yorkshire boats. The boats built at Riley Green are also not very well liked.'

Riley Green boats were lightly built, and consequently not really suitable for use in docks where they had to withstand being crushed by larger vessels. However, they were liked in the coal trade, where their fine lines and construction made them speedy, without reducing their carrying capacity. Other variations occurred with the introduction of steam and diesel

Tiger *was sold by Canal Transport Ltd in 1936, shortly after this photo was taken. Note the lack of scrollwork on the stern post, although there is some underneath the Liver bird. The mirrored style of scroll in the outer panel suggests a boatyard between Wigan and Liverpool, as does the decorative lettering, something which started to disappear from boatyards in the area around this time. The cabin was under the deck on which the boys are playing.*

powered boats when a reduction in living accommodation was the result of the installation of an engine.

On horse boats, there were cabins at bow and stern. That at the stern provided the main living accommodation, and was entered through a hatch, or scuttle hole on the left hand side of the deck, near to the hold. On all merchandise, and some coal boats, there was also a door in the rear bulkhead, below the scuttle, giving access to the hold. The headroom inside the cabin was only about four feet, and all furniture was built-in. A stove was fixed in the middle of the bulkhead, usually about twelve inches square by three feet high, and open fronted. A hob clipped on to the top of the firebars so that a kettle could be boiled, or food cooked. Some boats had an oven fitted to the side of the stove, but these were more unusual. Yorkshire stoves were also more ornate, with flowers and other decorations cast into the front plate. They were of tapering section, resulting in their distinctive square chimneys. Lancashire stoves finished with a flat top, curving up in the middle to a circular chimney. Chimneys were usually in two or three sections, those in Lancashire being linked by chain to prevent them being lost if they were knocked down.

On the right hand side of the cabin was the double bed, used by the captain and his wife, should she be accompanying him. It was separated from the rest of the cabin by a wall of six panels, three of which were hinged for access to the bed, though on some boats opening panels were not provided, and a curtain sufficed. At the rear of the cabin were three cupboards, the centre one was the largest, and it folded down to form a table. Two drawers were provided under the cupboards. There was a small shelf above the cupboards and bed panelling for ornaments. In the right-hand corner, against the bulkhead, was a locker, about eighteen inches square, and the full height of the cabin, which held the horse's proven. Finally, a bench was fitted against the bed panelling, under the table cupboard, and along the right hand side of the cabin. Coal, washing utensils, and other items were stored underneath. Footwear was stored under the bed, reached through a gap in the panelling under the bench.

The bow cabin was similar, the main difference being that a single 'bed hole' was provided on either side of the cabin. As the cabin was symmetrical, access was through a scuttle hole in the centre of the deck. With the introduction of engines, the stern cabin became the engine room, and the bow cabin the main living accommodation. The layout was unchanged, as few motor

*The **Naiad** was originally used for towing trains of boats. By the time of this photo near Rishton, c 1910, she had been demoted to moving dredging hoppers. Tugs had a simple paint scheme, with panels painted without the usual L&LC scalloped corner detail. Accommodation for the crew was in the bow.*

*Even into the late 1950s, boats continued to have some decoration. The **Timothy** of James Glover is seen here passing Apperley Bridge still painted with distinct panels and registration details.*

boats were used regularly by families, and accommodation was only required by the captain and mate.

Water was carried in a five gallon barrel, which rested on its side in a cradle. A dipper, like the old milk measures, was used, access being through a hole cut in the top of the barrel. The inside of the barrel was well covered with pitch to stop the water from becoming tainted. On deck, boats would often have a dog kennel, and a proven box was provided for the horse. This was filled each day from the store in the cabin. Merchandise boats were less likely to have a kennel, and their proven supply was usually carried in sacks draped over the sheets covering the rear of the hold. Ventilation of stern cabins was augmented by a hole, about nine inches by six inches, cut in the centre of the deck, which was covered by a small kennel shaped box.

It is difficult to say exactly how long it took to build a boat. The time varied with the number of men employed, and building could be interrupted by urgent repair work. Two to three months seems to have been usual, with two craftsmen, two labourers, and an apprentice employed full time. When the hull needed caulking, most other work at the yard would stop for a few days, and all available men would help.

The dimensions of the locks to which the short boats, working above Wigan, were built allowed an overall length of 62 feet, and a beam of 14 feet 4 inches. The long boats, which worked between Liverpool and Leigh, were 73 feet overall, with a beam of 14 feet 6 inches. A depth of 4 feet was guaranteed by the company, though the canal was deeper in Liverpool and at Leeds. The boats were built with a depth of 4 feet 8 inches centrally, measured from gunwhale to keel. The stem and stern posts were 6 feet high, resulting in a shear of 1 foot 4 inches,

***Irwell**, seen here passing Foulridge in 1939, was built at Northwich in 1934. Little in the way of decoration was applied to iron and steel motor boats, with paintwork just including their name, owner and registration.*

*Two photos of Richard Williams' **Margaret** in the 1950s; the stern view at Crooke and the bow view at Burscough. By this time, decoration had been reduced, with plain sans serif lettering, but the mirrored type of scroll still survives. This suggests that the decoration had been done at Parkes' yard in Liverpool or by Mayors', either at Wigan or by boatbuilders based at Tarleton using the drydock at the top of the locks at Burscough. (Edward Paget-Tomlinson)*

which improved the look of the boats. The exact measurements varied from yard to yard.

The original boats had fine lines, and it was estimated, in 1850, that the average load carried was 32 tons. By 1874, the canal had been deepened, and this resulted in an increase in the load carried to 38-40 tons. By the 1930s, the introduction of engines had made it less important to have a boat that could be pulled easily, as with horse boats. The shape was altered, enabling weights of around 50 tons to be carried by a single boat. The long boats, working on the Liverpool section, could carry about 20 tons more than that taken by a short boat.

Oak and pitch pine were used for construction. First the keel of 10 by 3 inch oak was laid, and stem and stern posts erected. Next the frames were placed on top of the keel. These were of oak, and where the frames were curved, would be cut from timber grown to shape. Each boat yard had their own set of patterns, from which the shape of the frames was taken. These dictated the final shape of the boat, which varied from yard to yard. The frames were spaced 18 inches apart, 3 inches wide and usually tapered from 4 inches to 6 inches. A keelson, originally of 10 inch square pitch pine, was fitted over the keel, and above the frames. Later an 8 inch by 6 inch steel joist was used. The bow and stern were strengthened by oak 'hooks', 12 inches by 4 inches fitted just below deck level from the stem post to the first cross beam, which also marked the start of the hold. Three 13 inch pitch pine planks were fitted either side of the keel, the rest of the planking being of oak, normally two inches thick, but increased to three inches for the rubbing strake, which protected the rest of the hull from abrasion. The decks were of 4 inch by 2 inch pitch pine, and 2 inch pitch pine boards were also used to line the hold. If a coal boat was required, two beams were fixed across the hold, and 9 inch wide side decks fitted. For a merchandise boat, the stern beam was made removable, so that long cargoes could be loaded more easily. The beams held the sides of the boat out, against the water pressure. The wide side decks, and coamings would then be fitted. The hull, after caulking, was pitched and tarred. It was then ready for launching. Cabins were usually fitted after the launch.

***Sir Robert**, the Skipton length spoon dredger, at Kildwick. The panels on the transom have more lining than Lancashire boats.*

*A detail from the photo of **Tiger**. From the scrolls and the slight serif to the letters, the paintwork could well have been done in East Lancashire. Certainly, patterns for the heart-shaped wooden pieces, used to protect the rest of the paintwork, survived at Whitebirk until its closure as a dock. The whole is typical of the canal company's boats, with no excessive decoration, but everything very smart and workmanlike. (The Boat Museum Archive)*

*Contemporary colour pictures are rare, so this postcard showing Ainscough's **Mary** at Burscough is particularly useful as the colour scheme matches the description given by old boatmen,*

Iron boats were first introduced in the 1870s during the shortage of craft following the termination of the railway lease. A specification was drawn up by Mr. Bartholomew, of the Aire & Calder Navigation, and six iron boats ordered from William Alsop, of Preston, in 1875, at a cost of £360 each. This was about £100 more than the equivalent wooden boat. Later, in 1891, two steel boats were purchased for the Leeds and Wigan traffics from Liverpool. Although iron was used for dredging equipment and tugs, no more iron or steel barges were ordered until the motor boats, built for Canal Transport Limited, were delivered in the 1930s, though the Wigan Coal & Iron Company, however, may have had some built in the 1890s.

Besides their higher initial cost, there were other reasons why iron boats did not become more widely used. None of the boatyards on the canal were equipped to repair them, and the necessary expertise had to be brought in from outside. They were disliked by the boatmen, as their weight and poor shape made them tiring for horses to tow, and the cabins were less comfortable than those on wooden boats.

No photographs of boat painters at work on boats on the Leeds & Liverpool Canal survive. The only contemporary illustration so far discovered is this view of the painter at British Waterways' Burnley yard, Finsley Gate, in the 1950s. By then he would mainly be producing signs as much of the boat repair work was carried out at Whitebirk yard.

A steam flyboat, probably ***Ophelia****, unloading grain at Skipton. Although there is limited decoration, the scrolls either side of 'Flyboat' and on the knighthead above the stem post are good examples, possibly from a dockyard in East Lancashire. The detail on the left is enlarged, but as with many photographs, the boat's decoration is but a small piece of the whole picture. Consequently, the scroll work often appears blurred and lacks definition.*

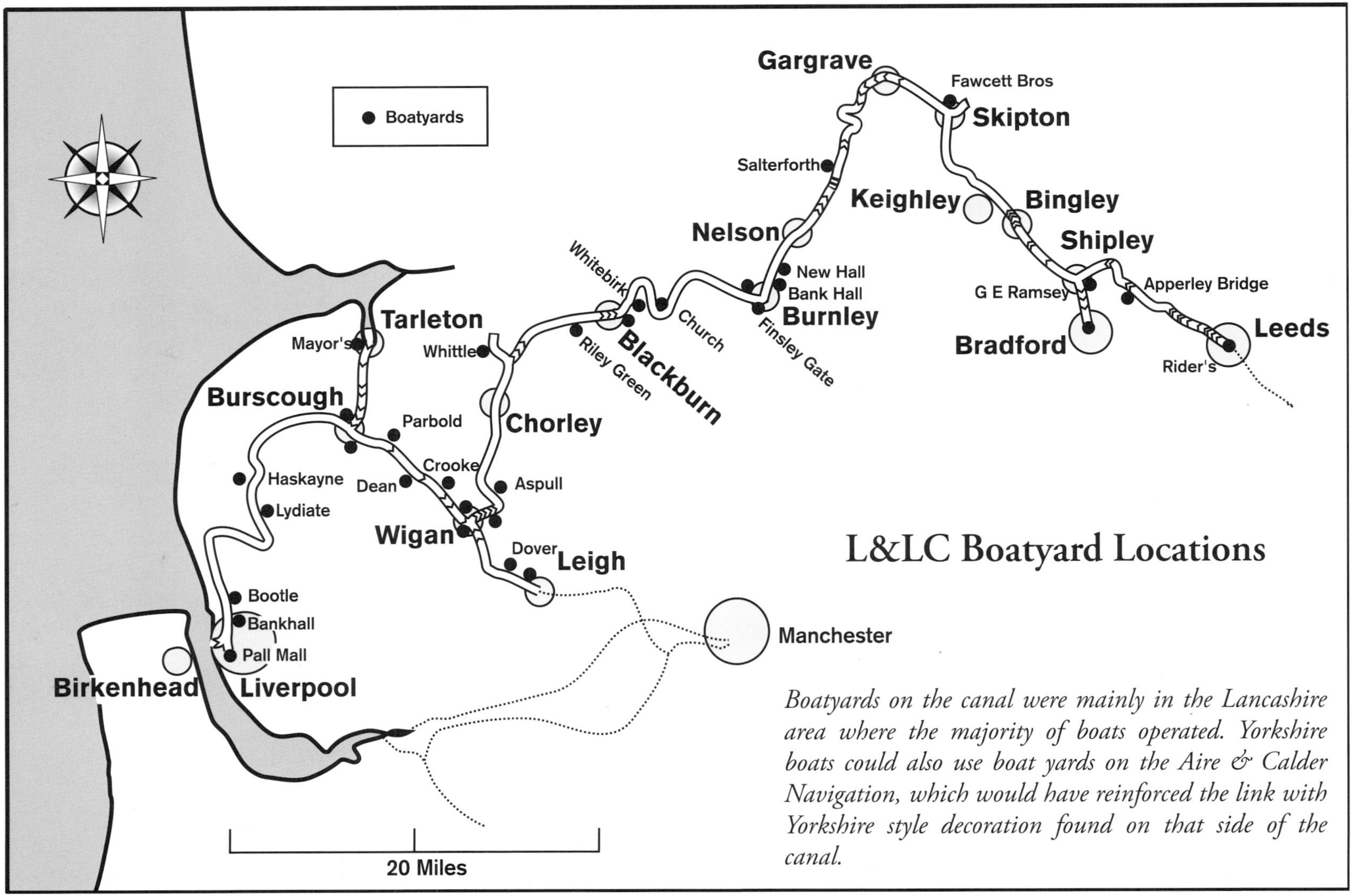

L&LC Boatyard Locations

Boatyards on the canal were mainly in the Lancashire area where the majority of boats operated. Yorkshire boats could also use boat yards on the Aire & Calder Navigation, which would have reinforced the link with Yorkshire style decoration found on that side of the canal.

Boatyards on the Leeds & Liverpool Canal

Liverpool basin: A drydock was available here until the basin was rebuilt during the enlargement of Exchange Station in the 1880s. Alongside was a wet dock used for gauging boats.
Bankhall: Slipway built by J. Parke & Sons Ltd. in 1934 after Pugh's at Bootle had closed. Using a small arm near Bankhall warehouse, it comprised a removable slipway for pulling boats out onto the bank. There were very limited facilities, the iron beam used for a kelson being bent by raising one end to the required height, weighting the beam and then building a fire around it, producing a gently upward curve at the beam end.
Bootle: J. G. Pugh Ltd. Side slip near Coffee House Bridge. Closed in the 1930s.
Lydiate: The Lunds had their yard with slipway here, operating well into the twentieth century. They docked many of the Richard Williams' boats and had a good name for decoration.
Haskayne: A boatyard is shown in an 1802 survey of the canal to the south-west of the bridge on the Wigan side of the Ship Inn.
Burscough: The drydock at the top of the Rufford line was probably built at the same time as the canal. In the early twentieth century it was used by J. Tyrer until his slipway opposite was opened, possibly in the mid-1920s. James Mayor used the drydock in the 1950s and '60s, particularly after giving up the Wigan yard.
Burscough: The slipway on the offside of the canal on the Wigan side of the junction with the Rufford line was owned by J. Tyrer and probably used up to the Second World War.
Tarleton: Side slip on site of the Douglas Navigation half-tide lock below the current tide lock; used by Mayors until c1910 when they moved to the present site above Tarleton Lock.
Parbold: Drydock on the outside of Parbold turn on the proposed main line of the canal. Early users could have been Cartmells, who built and repaired boats for Winstanley Collieries. Later operators were J. Sheldon who docked Richard Williams' boats. The yard closed in the late 1930s or early 1940s. Harry Leyland was the painter at closure, and he then moved to the Wells' dockyard at Stretford, working there until 1969.
Dean: Drydock shown on the 1802 canal survey on the site of

Mayor's yard at Tarleton in the 1930s, just after they had moved from their original yard on the River Douglas. Later another cover slip was built, the whole yard being located on the former railway interchange.

the present second lock.
Crooke: Boatyard on the level ground on the Liverpool side of Crooke Hall Inn. Closed following subsidence caused by mining.
Wigan: Ormandy & Co. repaired boats for Winstanley Collieries in the mid-19th century at a yard on Wallgate, near to Seven Stars Bridge.
Wigan: Slipway for two boats built in the 1880s during expansion of the canal company's fleet. The buildings alongside were used for maintaining the steam engines used in the boats. New boats built on the ground between the slips and the bottom lock. Kelsons were bent by a manually powered drop hammer of about 5 cwt in the company's maintenance yard on the towpath side of the canal. The yard was leased to Mayors from 1933 until the early 1950s. Bert Stretton did the painting here for Mayors, and was replaced by Peter Moss when he left.

Wigan yard, with the launch of ***Darlington*** *in 1951. There were two covered slips, the building berth closer to the lock, and the engine repair workshops in the middle distance.*

Wigan: Drydock above the bottom lock, probably built early 19th century.
Wigan: Double slipway above the bottom lock was built in the early 1950s as part of the maintenance depot, including lock gate production, for the whole canal. Closed c1980.
Dover: Possible sideslip operated by T. & W. Wells.
Leigh: Dock shown on 1894 OS map at Westleigh basin.
Wigan Locks: Boatyard for Ince Hall Coal & Cannel Co at lock 18. Probably closed in the 1870s when the company sold its fleet.
Aspull: Three slipways owned by Wigan Coal & Iron Company on the offside just passed the first bridge above Wigan Locks. There was also a timber yard and saw mills, with the office/house still standing.
Walton Sumit: Drydock shown on the 1850 OS map.
Riley Green: Covered and open side slips operated by J & J Crooke until c1963, later used for pleasure boat construction till c1980. Shown as a timber yard on the 1845 OS map, and continued as such alongside boat building.
Blackburn: Drydock, later converted to wet dock and site now covered by Cobble factory. Evidence of entrance in wall on offside.
Whitebirk: Covered sideslip and two open slips operated by T & J Hodson from c1880 to c1963. Site demolished April 1983.
Church: Boatyard on Leeds side of Church swing bridge, closed c1870.
Burnley: Dockyard shown on 1845 OS map on offside near Burnley Barracks. Probably closed c1920.

The slip at Salterforth, here around 1939, was very much a country business, with farming and quarry repairs also undertaken.

Burnley: Finsley Gate slipway used by the canal company for maintenance, but may have originally been used by John Birley.
Burnley: Drydock at Bank Hall originally worked by Birleys from early 19th century. Site developed when Bank Hall Colliery built, and closed c1963. Towards the end it was used by Hodson's of Whitebirk as needed.
Burnley: Dockyard shown on Leeds side of Colne Road Bridge until mid-19th century.
Burnley: New Hall dockyard shown on 1894 OS maps to the north of Daneshouse.
Salterforth: Sideslip 200 yards on Liverpool side of Salterforth Bridge. Also undertook agricultural repairs.
Skipton: Slipway at junction with Springs Branch used in nineteenth century and operated by Fawcett Bros until the late nineteenth century.
Shipley: Drydock and slipway on Leeds side of junction with the Bradford Canal. Operated by G. E. Ramsey in 20th century.

Apperley Bridge: Sideslip and dock on offside below swing bridge operated by G. E. Ramsey around 1910, Thomas Avery in 1927 and Canal Carriers Ltd of Shipley from 1930s to '50s.
Leeds: Two drydocks with boat building area between docks and river. Operated by W. Ryder & Co., who were also timber merchants, from the 1930s.

Rider's yard at Leeds was almost alongside River Lock, right in the centre of Leeds behind the railway station.

Notes on painting boats at Whitebirk

Preparation: no sand paper, masking tape, etc., just scrapers fashioned in the smithy fire and made from flat files. Sometimes 'scrunting', pieces of broken glass, would be used. A wire brush and chipping hammer were used for any metal plates. A caulking iron would often be used as a putty knife.

Brightwork: two coats of paint and then one coat of varnish. After the first coat, putty up using a linseed oil putty prior to the second coat. On scraped timber, two coats of varnish were used.

Decks: two coats of paint, the first coat 'scrubbed' well in using a well-worn brush. All the 'best' red surfaces and the engine cabin sides received one coat of varnish.

Brushes: One large 'rigger' was needed for yellow edging lines, and two or three sable brushes: a duck (no 6), goose (no 8), and a swan (no 10). The names originated from the size of the quill once used to make brushes, a lark being minute. Writing brushes, when not in use, were kept in shape using a little slip-grease, supplied for the slips on which boats slid into the water after repair. The brushes were thoroughly cleaned before use. For more general painting, the brushes were two or three inch, with light and dark colour brushes kept separate, hung with their bristles in tins of water to stop them from drying. A two inch oval brush was used for varnish, and this was kept suspended in a container of white spirit. The brushes used for black varnish were also kept separate in their own tin of water, while the shellac-knotting brush would be cleaned out with methylated spirit after use.

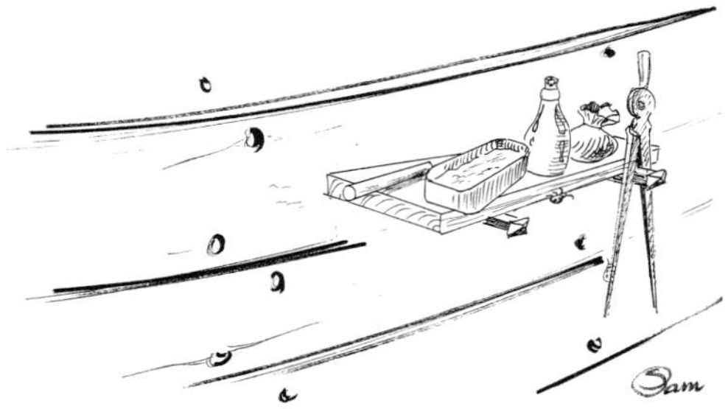

A temporary work bench could be fashioned from two spikes hammered gently into the planking.

Paints: red lead came in powder form, with stainers bound in oil for the colours — yellow ochre, Brunswick blue, Brunswick green, umber, lemon chrome and 'best' red — and white lead, etc, coming in paste form. The stainers came in half hundredweight or 28 lb drums and had a consistancy similar to butter. They were covered with water to prevent skimming over. Raw linseed oil (Riley Green yard used boiled) and sun dryers were added to make the required colour. Drop or Berlin black came semi-mixed. White used for scrollwork etc had a mere morsal of blue added. This counteracted the yellowing of the linseed oil base. Slick was the name given to a mixture of best red and varnish which was used on bollards, etc. In earlier days, the paint was ground at the boatyard, with putty made up from whiting and raw linseed oil: *'Putty and paint are a devil of a saint!'* Putty was kept in a galvanized bucket, with around half a hundredweight made up. The bucket had to have a lid

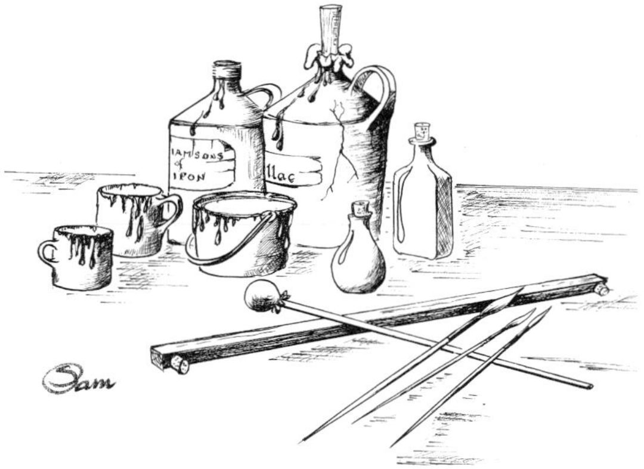

Signwriting tools used at Whitebirk.

to prevent mice from eating the putty. If they did, there were white droppings everywhere! The tins in which the paint was mixed were burnt out afterwards for re-use, and lumps could be removed using a seive made from a can with holes punched in the bottom. An area where painting or varnishing had been skimped was called a 'holiday'. American turps (ie. distilled in America) was usually used for sign writing. The use of emerald green had to be recorded in the poison register.

Painting: In frosty weather, hot iron plates were used to warm areas prior to painting. On summer mornings, willow herb floating on the wind was a problem, while varnishing was always left till the late afternoon when there was less dust around and less possibility of blooming caused by dampness. 'Sessing' was caused by an oily surface and could be cured by breathing on the surface. Another problem was 'crocodiling' where the surface became crazed. To comply with health regulations, the cabins were usually painted around every two and a half years.

Sign writing equipment: two maulsticks (or mahlsticks) for steadying the hand on detail work, as was the straight edge with cork at either end to raise it off the paint, a ponce bag filled with chalk to dab onto a stencil, chalk and a chalk line, rule, dividers, a bottle of American turps, a duster brush, three sable brushes: a crow (no 7), a swan (no 10) and a large rigger.

Modern colours: The modern approximate Pantone equivalent of the colours used are as follows:

Bull red, 186; Umber, 161; Yellow, Lemon chrome; Light blue, 279; Dark blue, 280; Light green, 376; Mid green, 363; Common red (red lead), 151.

However, originally the colours would vary slightly from mix to mix, with special colours sometimes mixed for special purposes.

The paint cupboard doors were covered in paint from knocking out brushes. The weight of the lead-based paint caused the hinges to collapse, and new ones were added on several occasions.

Glossary of boat terms

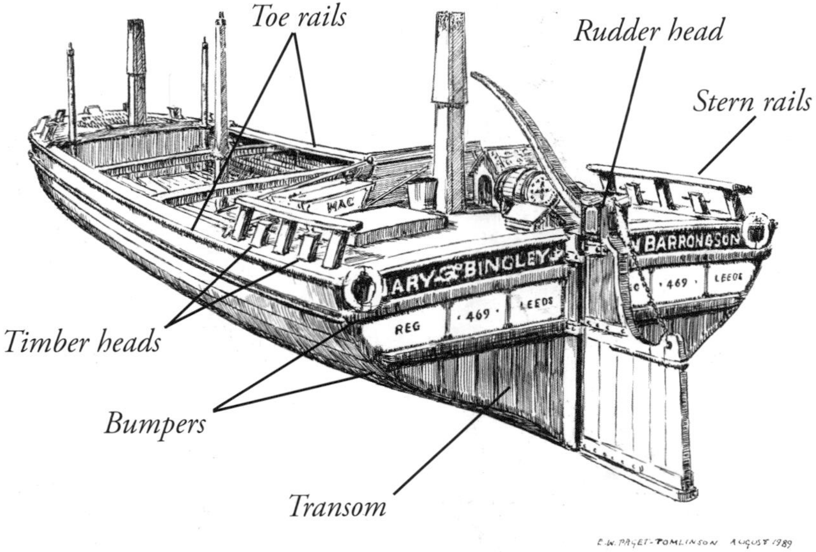

Bumper: protruding wooden strips at bow or stern which protected the boat from damage in locks and elsewhere.

D-iron: iron strip, rounded on one side, which was fitted to the edge of planks to prevent wear.

Dumb boat: an unpowered or horse-drawn boat.

Flyboat: On the Leeds & Liverpool Canal, a canal company boat carrying general cargo and working non-stop or to a timetable.

Knighthead: wooden frames either side of the stem post, originally used to help support the mast rigging on sailing boats. They were unnecessary on canal boats and ceased being fitted around 1900.

Toe rail or **wash strips**: a small raised wooden strip fitted to deck planks about one inch in from the edge of the boat. Often painted red.

Transom: the square stern often used on Leeds & Liverpool Canal boats.

Further reading

Wheat, Geoff, ***Leeds & Liverpool Canal Craft***, 1972 — Includes a possible colour scheme for B C Walls' boats.

Lewery, Tony, 'The Northern Tradition', ***Waterways World***, June 1975.

http://www.mikeclarke.myzen.co.uk/LandL.htm has a link to a pdf with brief details of traditional L&LC paintwork.